Why Do We Go to the Zoo?

The Fairleigh Dickinson University Press Series in Communication Studies

General Editor: Gary Radford, Department of Communication Studies, Fairleigh Dickinson University, Madison, New Jersey.

The Fairleigh Dickinson University Press Series in Communication Studies publishes scholarly works in communication theory, practice, history, and culture.

Recent Publications in Communication Studies

Patricia Arneson, *Communicative Engagement and Social Liberation: Justice Will Be Made* (2014)
Erik A. Garrett, *Why Do We Go to the Zoo?: Communication, Animals, and the Cultural-Historical Experience of Zoos* (2013)
Philip Dalton and Eric Kramer, *Coarseness in U.S. Public Communication* (2012)
Catherine Creede, Beth Fisher-Yoshida, and Placida Gallegos (eds.), *The Reflective, Facilitative, and Interpretive Practices of the Coordinated Management of Meaning* (2012)
Jolanta Aritz and Robyn C. Walker, *Discourse Perspectives on Organizational Communication* (2011)
S. Alyssa Groom and J. M. H. Fritz, *Communication Ethics and Crisis: Negotiating Differences in Public and Private Spheres* (2011)
R. C. MacDougall, *Digination: Identity, Organization, and Public Life* (2011)
Deborah Eicher-Catt and Isaac E. Catt (eds.), *Communicology: The New Science of Embodied Discourse* (2010)
Dan Cassino and Yesamin Besen-Cassino, *Consuming Politics: Jon Stewart, Branding, and the Youth Vote in America* (2009)

On the Web at http://www.fdu.edu/fdupress

Why Do We Go to the Zoo?

Communication, Animals, and the Cultural-Historical Experience of Zoos

Erik A. Garrett

FAIRLEIGH DICKINSON UNIVERSITY PRESS
Madison • Teaneck

Published by Fairleigh Dickinson University Press
Copublished with Rowman & Littlefield
4501 Forbes Boulevard, Suite 200, Lanham, Maryland 20706
www.rowman.com

10 Thornbury Road, Plymouth PL6 7PP, United Kingdom

British Library Cataloguing in Publication Information Available

Library of Congress Cataloging-in-Publication Data
Garrett, Erik A.
Why do we go to the zoo? : communication, animals, and the cultural-historical experience of zoos / Erik A. Garrett.
p. cm. — (The Fairleigh Dickinson University Press series in communication studies)
Includes bibliographical references and index.
ISBN 978-1-61147-645-3 (cloth : alkaline paper) — ISBN 978-1-61147-646-0 (electronic)
ISBN 978-1-61147-870-9 (pbk : alk. paper)
1. Zoos—Social aspects. 2. Zoos—Psychological aspects. 3. Zoos—Philosophy. 4. Zoo visitors. 5. Zoo animals. 6. Human-animal communication. I. Title.
QL76.G37 2014
590.73—dc23
2013032415

∞™ The paper used in this publication meets the minimum requirements of American National Standard for Information Sciences Permanence of Paper for Printed Library Materials, ANSI/NISO Z39.48-1992.

Printed in the United States of America

For Lara and Konrad, and a future full of furry and feathered friends

Chuang Tzu and Hui Tzu
Were crossing Hao River
By the Dam.

Chuang said:
"See how free
The fishes leap and dart:
That is their happiness."

Hui replied:
"Since you are not a fish
How do you know
What makes fishes happy?"

Chuang said:
"Since you are not I
How can you possibly know
That I do not know
What makes fishes happy?"

. . . .

"I know the joy of fishes
In the river
Through my own joy, as I go walking
Along the same river."

—Chuang Tzu, "The Joy of Fishes"

Contents

Acknowledgments

This is my first book and I owe a great debt of gratitude to many fellow travelers who helped me along the way. First, I want to thank my wife Patrycja for her enduring support, patience, and love; my children Lara and Konrad for their love, putting up with their philosopher father, sharing their "multi-animaled" world with their Dad, and for our many trips to zoos; my sister Tatiana for her passion; Dane; and my parents James and Ilona for teaching me compassion by raising me in an "originary" world full of animal Others. My mother especially served as a role model of how to interact with both the world of animals and ideas. Also thanks for the additional editing. Another set of thanks goes to Babcia for providing me the freedom and time to finish writing this book.

I want to thank my fellow travelers at the Communication Research Department of Brookfield Zoo and the wonderful scholars I interacted with during my time there. In particular I want to thank Carol Saunders, Todd Gieseke, Andrej Birjulin, Gene Myers, Peter Kahn, Jim Cantrill, Jill, Janine, Jean, and of course Paul Walenga.

Also I wish to thank my colleagues at Duquesne University and the Department of Communication & Rhetorical Studies for support. The department was generous in granting me excellent research assistants such as Lindsay Karr, and especially Kelly Cummings who helped me tremendously. I want to also thank all my students and colleagues at Duquesne University for the dialogue. I also want to thank Fairleigh Dickinson Press for a smooth review process. Thank you to Gary Radford, Harry Keyishian, and the two anonymous reviewers for very helpful comments.

Finally, I want to thank all my various teachers. In particular, I want to thank Marilyn Nissim-Sabat and Lewis Gordon for teaching me phenomenology. You both have provided me with excellent models for how to be scholars with a soul. You are two of the smartest people I have ever met and yet have always been generous to me and countless others. Also, thank you for helping carve out the Phenomenology Roundtable for me to learn to engage in philosophical conversation in a supportive environment that gave me intellectual brothers and sisters like Mike Michau, Mike Monahan, Carolyn Cusick, David Fryer, Ken Knies, Doug Ficek, and Neal Rosen. I also want to mention particularly important comrades at Purdue University like Devika Chawla, Katerina Tsetsura, Robyn Remke, Cory Williams, Sol Neely, Brian Kanouse, Shane Wahl, Sharron Hope, Chris Darr, Joel "Pookie" Krueger, Lee McBride, Kristen Lucas, Jacoby Carter, Damion Waymer, Ada, Ron, Olga, Anna, Mariya, Bertha, Lazar, Rebecca, Winship, and the many others who I could name that made Lafayette, Indiana, bearable.

I also want to thank my wonderful teachers at Purdue University such as Bill Rawlins, Don Burks, Josh Boyd, Melanie Morgan, Jane Natt, Sandor Goodhart, Alan Beck, Amardo Rodriguez, Austin Babrow, Sister Ann Astell, and the great Continental philosophers who taught me so much: Calvin O. Schrag, Dan Smith, Tom Ryba, Leonard Harris, Charlene Seigfried, and Martin Matustik. This project was inspired by two chapters from my dissertation. Therefore, I need to thank my committee members Jennifer Zeigler and Mark Bernstein for their assistance. But in particular, the most thanks needs to go to Felicia Roberts and Bill McBride who spent countless hours with this project. I was very lucky to find not only smart thinkers, but good human beings as well—thank you both. You each helped me on this journey. Felicia shared a love of animals, and helped me find a home in a discipline I still often feel like an outsider. The best advice I received at graduate school was to work with Bill McBride. No one fights as passionately for his students and he is a world-class philosopher. I try to provide the helpful focus and nudging that you gave me to my graduate students pursuing and developing their own ideas.

Thank you all!

Knutmania

More people visit zoos and aquariums in the United States each year than attend all sporting events combined. In 2010 alone, more than 178 million people visited one of the Association of Zoos and Aquariums' (AZA) 212 accredited facilities.[1] (The AZA was formerly known as the American Association of Zoos and Aquariums, so most of its members are in the United States.) According to the World Association of Zoos and Aquariums (WAZA) approximately 700 million people visited one of the world's approximately 10,000 zoos. This is more than ten percent of the world's population.[2] This is an amazing statistic that underscores the popularity of zoos and aquariums and the large role they play in our everyday lives. Yet, despite the hundreds of millions of visitors each year, these spaces have remained outside of the realm of philosophical analysis, or for that matter, most liberal arts fields' domain of investigation. This lack of theoretical examination is also interesting considering the paradoxical position that a zoo is situated in, being a space of animal confinement and a site that provides valuable tools for species conservation and public education.

This is a book about zoos that is meant to look at the experience of zoos in a new way. Following in the footsteps of Edmund Husserl's transcendental phenomenological analysis, this book examines the experience and the context of our relationship with animals in the zoo setting. In particular, I am interested in exploring what makes this such an important experience for children, since this indicates powerful early bonds with the animal world that will influence us later in life. I also am giving this phenomenological method my own spin that focuses on the lived communicative experiences in context. This method can be thought of as a "*praxis phenomenology*" since it seeks to do phenomenology rather than merely explicate meanings. Ultimately, this is a book that seeks to answer the question of why we go to the zoo. I w with an important pop cultural reference to a zoo animal and a story from my past to contextualize how wide and deep an impact the zoo animal bond has in our lives.

KNUT

Millions of people from around the world watched a tiny, fluffy, and furry ball of white fuzz on YouTube. The baby polar bear was less than a foot long and had fur that looked as white as the Arctic snow. He had big round eyes, a tiny button nose, and paws that seemed too big for his frame; hinting at the size that was yet to come. This was arguably one of the most photogenic animals of all time.

One of the videos showed a polar bear cub only a few weeks old lying across the forearm of a zookeeper while being combed. As the comb rhythmically stroked his fur, the little bear stuck out his tongue. There was playfulness in the gesture. We then saw the keeper feeding the polar bear from a baby bottle and the spirited interactions between the tall brown-haired German keeper and the polar bear cub. They play-fought and wrestled. In every photograph and video clip the cub seemed to get older as he climbed on the keeper.

The baby polar bear, Knut, was born in Germany's Berlin Zoo on December 5, 2006. It was the first polar bear born in the zoo in more than thirty years. Knut and his twin were rejected by their mother at birth. Tosca, their mother, was a former circus animal that had worked in East Germany. For whatever reason, she left the two cubs alone on the rock at the edge of her enclosure and showed little to no interest in her own two offspring. The zookeepers quickly removed the two small cubs that were not in very good condition. Knut's unnamed sibling died and Knut himself needed an incubator for the first month and a half of his life. Knut weighed less than two pounds at his birth. After that, he was "hand-raised" by a zookeeper named Thomas Dörflein. Mr. Dörflein moved into the exhibit and stayed with Knut. The videos turned Knut into a media sensation, so there was much attention and anticipation before his formal debut to the public at the zoo on March 23, 2007. Over 400 journalists from around the world awaited the chance to get their own pictures of the adorable Knut.[3] Eventually this beautiful animal would appear in an Annie Leibovitz photomontage with Leonardo DiCaprio on the April 2007 *Vanity Fair* cover.[4] You know you have made it when you get equal billing to DiCaprio in a photo shoot!

As "Knutmania" reached its peak in March 2007, a German animal rights activist, named Frank Albrecht, made quite a sensation when he suggested that it would be better to have Knut "put down" rather than have him raised by hand at the zoo. Albrecht's argument was that Knut has certain inherent and inalienable rights. Among those rights is freedom. "Hand-raising" the little polar bear would make him too reliant on humans and therefore take away his independence and ability to be free. Therefore, Albrecht and a small number of animal rights activists suggested the alternative of euthanasia instead of letting the little guy continue living in such an "inappropriate" environment. To these activists, the zoo's saving Knut had violated the laws of "nature."

But Knut was never in danger of being put down by the zoo. "'We never thought to put him to death or let him die,"[5] said Berlin Zoo veterinarian Dr. Andreas Ochs. People from all around the world wanted to see the famous polar bear. In 2007 alone, Knut brought in an increase of 5 million Euros in tickets and merchandise for the Berlin Zoo.[6] There were Knut T-shirts, stuffed animals, books, figurines, and even a Knut movie. Knut literally had legions of fans from around the world that flocked to see him, with as many as 15,000[7] in a single day at the peak of his popularity. Before there could be the legions of fans there had to be the photographs and images of the playful and seemingly mischievous bear romping around. Media outlets from across the globe reported on major events in his life and Internet videos earned millions of hits. Knut was a celebrity bear and the star of the Berlin Zoo.

Yet, tragedy was a frequent part of Knut's life. In September 2008, Thomas Dörflein, who had hand-raised Knut from his birth, died of a sudden heart attack. It was the charming closeness and playfulness of the interactions between the tiny Knut and the big-bearded German man that warmed the hearts of many onlookers. There was a bear-like appearance to Dörflein, who would give the tiny cub a bottle and snuggle with the white ball of fur. Mr. Dörflein accepted Knut, too. He moved into the enclosure since Knut required feedings every two hours. He accepted the scratches and bites of the playful Knut that grew stronger and more dangerous with each passing week. After the death of the person on whom he had most closely imprinted, there were attempts to provide Knut with companionship of other polar bears. The Berlin Zoo had differing levels of success introducing him to other female bears. Polar bears tend to be solitary animals in the wild and there were complaints and concern about socializing Knut with other polar bears and the stress it may or may not have caused him. The zoo even reintroduced him to the mother who had abandoned him.

As he matured Knut displayed behaviors that zookeepers didn't observe as frequently, if at all, with the other polar bears in the zoo. Unlike the other polar bears at the Berlin Zoo, Knut would often stand on his hind legs and appear to "wave" at his audience. Visitors reported that he even "played with them," describing numerous examples of Knut tossing the ball into his moat inviting the guests to play catch.[8] Four years after his debut Knut continued to attract visitors.

However, the playful nature of the world-famous bear started to slowly change. Knut began to display a phenomenon known in the zoo world as "stereotypic behavior,"[9] or "abnormal repetitive behavior."[10] Knut like many other polar bears in captivity began pacing in a regular routine pattern. This obsessive-compulsive behavior is common in many large animals kept in captivity. Scientists and zoos have spent millions of dollars studying the phenomenon and trying to decrease its occurrence. On March 19, 2011, before a crowd of approximately 600 to 700 people, Knut began repeatedly circling on top of a rock in his enclosure. His leg started to shake and he began staggering. Suddenly Knut fell into the pool in his enclosure and died. Just like every moment of his life, this was also captured on video and placed on the web garnering hundreds of thousands of macabre hits. A necropsy (animal autopsy) revealed that although the immediate cause of death was "drowning," Knut had been suffering from encephalitis, or swelling of the brain. This was most likely caused by an infection that Knut could have been suffering from for quite a while. The prognosis was that he would not have survived much longer even if he had not drowned.

Memorials poured in from around the world. People grieved online as if they had lost a member of their family. It is common practice in zoos that when famous or large animals die that they are often turned over to local museums of natural history to join their taxidermy collections. The thought of this being done to the "beloved" Knut caused outrage and protest. His remains were sent to the Berlin Natural History Museum to be displayed along such other famous deceased Berlin Zoo animals, as Bobby the gorilla (the first gorilla displayed at the Berlin Zoo), and the hippopotamus Knautschke (who survived an injury during World War II to become the star of the zoo during his era). Nearly 10,000 people signed an online protest against Knut's display in the Natural History Museum. One comment of protest came from Anna Schenk: "It was decided that 'our' Knut will not find peace after his death. We cannot, and will not, accept that."[11]

It was argued that Knut should not have been allowed to live from birth and that his hand-rearing by humans in captivity was inherently evil and harmful. Yet, Knut transcended the captivity of the zoo and affected so many people around the world. What was it that people identified with? What made Knut a star? What brought the crowds and legions of fans in the flesh and on the Internet? What was so special about this bear? Why did his death cause such an outpouring of sadness? Why did we visit Knut in the first place? And what makes us go to the zoo anyway?

GROWING UP AT THE ZOO

For the first fourteen years of my life I lived near Brookfield Zoo outside of Chicago. Some of my earliest memories are of my mother walking with my sister and me to see the animals. The zoo even predates my "active" memory. A story of many visits with my grandmother at the end of her life involved a recounting of how I would love to take her to the Reptile House to see the snakes over and over while my mother was delivering my sister. Since I heard her repeat this story countless times, I cannot be sure how much of my memory of Brookfield Zoo's Reptile House in 1979 was part of my "real" three-year old experience and how much of the memory is a reconstruction through the eyes of my grandmother's narrative. Either way, there is an impression upon my memory of walking hand-in-hand with my grandmother past the crocodiles and various snakes as she feigned fear.

However, I am very sure of other memories from the zoo. I remember that my favorite animal was Olga the walrus. I used to love to see her "show" on the south side of the zoo. Olga would blow kisses to the crowd and spray water at the children at the prompting of the keepers. The enclosure of the old walrus was always the first place I would want to visit. I would be very disappointed if Olga was not in view. I distinctly remember the sadness when she died in 1988 having attained the status of the oldest age ever recorded of a living walrus. My twelve-year-old sadness was very similar to the descriptions and memorials I observed for Knut. It was as if a close friend or family member had just passed away.

One of my first jobs right out of college was working in the zoo. With my training in philosophy, communication, and qualitative method analysis, my job was working in the Communication Research Department. At the time, Brookfield Zoo was one of only three zoos in the United States with a communication research department. We provided communication research and analysis for a wide number of goals throughout the zoo. However, when asked what my job exactly was, I would say, "I *watch the people* as they watched the animals in the zoo." I spent hundreds of hours silently observing interactions of "visitors" throughout the zoo. I interviewed countless numbers of people both formally and informally as they went about their zoo visit. I poured over thousands of pages of quantitative data from surveys visitors filled out.

The communication research department was led by Dr. Carol Saunders, and was a pet project of Dr. George Rabb, the director of Brookfield Zoo during my time there. The work of this group was grounded in "conservation psychology." As a matter of fact, many of the early studies we conducted were some of the first published in this area.[12] In the past twelve years there have been books published on conservation psychology, special issues of journals devoted to the topic, entire conferences devoted to it as well, and even a special session at the American Psychological Association. A detailed history of conservation psychology can be found on the website, www.conservationpsychology.org. However, one of the driving questions that motivated the research was what made someone care about nature and what role, if any, could zoos play in forming and maintaining these bonds? The experience of visiting the zoo is what I wish to explicate in this book. Why is the annual attendance of zoos and aquariums higher than attendance at professional sporting events around the globe? Why do we go? What do we get out of it? I hope to bring my various professional and personal experiences to bear on this experience.

After working at the zoo, I worked at the Center for the Human-Animal Bond at Purdue University. My dissertation investigated the bond between children and animals; I explored the effects of different settings and metaphors on the relationship. Now I am a father of two children of my own, living in Pittsburgh instead of Chicago, and I continue to make the zoo visit a part of the childhood tradition of my family. My situation is not unique. A majority of zoo visitors are families. What are my children, or any child, getting from this experience?

FOR MY READER

This project has a twofold purpose. First, it hopes to shed some light into the questions: Why do we go to the zoo? What is the nature of the communication, relationships, and bonds that are formed in this paradoxical setting of captivity and wilderness. Many of the visitors are family members, or school groups with children, so what possible developmental benefits are obtained by children's trips to the zoo? Surprisingly, zoos have not received the in-depth attention they deserve.

A secondary purpose of this study is to provide a methodological example of the doing of phenomenology. Having been trained as a Husserlian phenomenologist, let me state how difficult it is to find clear guides into the process of applying phenomenology to particular problems and areas of investigation in various disciplines. The praxis methodology to be used to investigate the experience of zoos in this book is the philosophical school of phenomenology in connection with rhetorical studies. In addition to answering the question, "Why zoos?" I hope my phenomenological methodological attention will be useful to others thinking of "doing phenomenology" as well. A secondary goal is to provide the phenomenologist with a practical praxis-oriented investigations with this methodological philosophy that has at times been summed up by the cry, "to the things themselves," which supplies the thinker with a powerful mode of investigating lived experience.

When I began my formal studies in phenomenology as a graduate student I was surprised that there were very few "how to" books on phenomenology. Most recent approaches were written by philosophers debating the interpretation, meaning, or context of an important theoretical precept of phenomenology. A majority of the work was exegesis of text and interpretation of Husserl, Heidegger, and their followers. As a philosopher I appreciate this work. However, I also have my foot in the discipline of communication. Communication is a relatively new formal discipline, only about one hundred years old.[13] The discipline of communication brings together quantitative social scientists and qualitative humanities scholars investigating that most interesting of human phenomenon—communication. So my interest in phenomenology from the communication side is less concerned with exegesis than it is with actually "doing phenomenology." I also believe this "living" approach of "doing phenomenology" is closer to the type of work that the founders of phenomenology—Husserl, Heidegger, Merleau-Ponty, and Sartre—were actually doing. However, phenomenology is a philosophical method of investigation for examining the structures of experience in a very practical manner. Phenomenology should be very valuable for communication researchers.

In the field of communication, the term phenomenology has often been abused. It is commonly used merely as a synonym for experience. Rather than shame individual authors, I will allow those readers interested in confirming this hypothesis do their own key word search of the term "phenomenology" in a search engine that serves communication scholarship. I would rather draw attention to those who have done foundational work in philosophy and communication and had an excellent grounding in phenomenology. Traditionally there seems to have been two schools of philosophy of communication coming out of Southern Illinois University and Purdue University that seem to have historically had a rich understanding of phenomenology. At Purdue, where I studied, this was caused by the insistence that graduate students regularly take classes outside of our discipline. Many students took classes in the philosophy department and a close bond was formed. In particular, we owe Calvin Schrag a great deal of gratitude since he is a major figure in twentieth-century philosophy who took the time to engage both the issues and the discipline of communication. Dr. Schrag published in our communication journals, wrote books on issues of communication, like *Communicative Praxis and the Space of Subjectivity*, and taught other great phenomenologists like Michael Hyde, Ramsey Eric Ramsey, Bryan Crable, and Corey Anton. At Southern Illinois University, the Department of Communication hired two phenomenologists, Lenore Langsdorf and Richard Lanigan. Both Drs. Lanigan and Langsdorf have written the seminal works in philosophy of communication from a phenomenological perspective. Their work has received and continues to receive the attention of philosophers, communication scholars, and those in other disciplines. At Duquesne University, where I currently teach, we are working to establish the same tradition within our graduate program.

I consider this work to build on the great phenomenological scholarship being done by scholars of communication. Yet, I also want to say that I am taking this in a new direction. I consider the use of phenomenology in this book to be a *praxis phenomenology*. In a sense all phenomenology is "applied," since it is directed out to the world. However, this book puts forth a particular phenomenologically focused method that is attentive to the communicative structures of our manifold experiences. Praxis phenomenology is differentiated from other forms of phenomenology in that it moves away from exegesis and attempts to *do* phenomenology in an active sense. In a sense, I consider this method more of a turning back to the praxis-oriented work done by both Husserl and Merleau-Ponty. Praxis phenomenology is an attempt to provide an examination of how particular experiences in specific settings are constituted as meaningful.

I believe the zoo will provide an even more important and necessary role in the child-animal bond as the places for "wild" encounters with animals and nature disappear and we continue the trend of becoming more urban in our lifestyles across the globe. Zoo animals allow us to experience the wondrous animals of the wild, but in a different setting that does not allow for the full potential display of their nature. The future status of the zoo is even more in question with cases of zoo animal attacks (like the big cat mauling in San Francisco), zoo type animals found in the home (like the recent chimpanzee face attack), and increasing celebrity status of animals (like Knut the polar bear from Berlin) that hit the media cycle. I also hope that this project will make a contribution to the new burgeoning division of environmental communication and environmental rhetoric within the field of communication. Explicit writing on environmental communication can be traced back to at least the 1980s with case studies and rhetorical approaches[14] focusing on how we communicate about the environment. However, as environmental communication research has expanded beyond a few rhetorical research projects it has become more multidisciplinary and has embraced a number of different approaches. Now many of the professional organizations of communication have divisions focusing on environmental research, like the International Communication Association (ICA)'s International Environmental Communication Association which has published the journal *Environmental Communication: A Journal of Nature and Culture* since 2007. For those interested in an introduction to environmental communication I highly recommend Robert Cox's *Environmental Communication and the Public Sphere*,[15] which does a nice job situating environmental communication within larger public policy debates and Craig Waddell's *Landmark Essays: On Rhetoric and the Environment*,[16] which contains most of the primary early articles in the field.

Zoos are a popular and common part of the lives of many humans across the planet. However, despite the definite rhetorical and philosophical issues with which they are struggling, for the most part, the work of zoos has remained outside of the gaze of the humanities. I believe that they have been ignored either because they are viewed as insignificant, or because of the reductive presuppositions about a zoo's—fallibility that it is merely a place for keeping animals in confinement. As we engage the phenomenological method, one of the first steps is a bracketing of prejudices. This project begins from the space created by this phenomenological maneuver. It is important to distinguish that the question of starting points separates this project from other rhetorical, philosophical, and ethical writings on zoos that begin with a set of views about the status of zoos in the world in relation to a set of beliefs. I will differentiate this book from other philosophical writings on zoos below, but for now I want to differentiate that this book is not written in the "animal rights" tradition that critiques zoos from the perspective that animals have certain inalienable rights which are denied by their confinement in zoos. The most famous examples of this tradition are Peter Singer[17] and Thomas Regan.[18] The experience of zoos is organic, in the original sense of the word, meaning that the prereflective experiences in which the zoo is entered in by the average person does not take place within the context of utilitarianism (Singer), or other moral claims as to whether a zoo is good, bad, just, or evil. If you are interested in a fresh approach to the experience of zoos this book is for you.

I want to acknowledge the work done by Stephen Spotte. His book, *Zoos in Postmodernism: Signs and Simulation*,[19] operates along a vein similar to this book. His perspective is writing as a curator quite expertly applying the methodology of semiotics to his analysis. This is an excellent examination of the status of the modern zoo by an insider. Also, as someone who dabbles in semiotics, his application of semiotics makes the reader smarter. When I sent out the proposal for my book, I was actually scared when it was suggested I read Spotte's book. In some ways we have a very similar goal. However, we use completely different methodologies and he takes a much more critical stance then I do.

Also, I want to acknowledge one other attempt to use phenomenology to analyze a zoo. In his essay, "Through a Frame Darkly: A Phenomenological Critique of Zoos," the philosopher Bernard E. Rollin attempts a phenomenological explication of the zoo experience. While I admire his work, I do not believe that his description is indicative of the zoo experience and is more likely a symptom of the clinical depression he admitted to feeling during the time of his zoo visits, "though I never got a formal diagnosis, I was seriously depressed and physically debilitated."[20] However, my main disappointment was in the failure of the essay to deliver phenomenology. As a matter of fact, the only phenomenological concept used in this essay is Heidegger's "ready at hand."[21] Unfortunately this translation is problematic because it confuses two important technical terms in Heidegger's phenomenology which actually may have the opposite connotation to how Rollin uses them. Heidegger makes use of both the terms *Vorhandenheit*, present-at-handedness, and *Zuhandenheit*, ready-to-handedness. In *Being and Time*,[22] these two terms are presented as various modes of encountering objects in the world. *Zuhandenheit* involves the disappearance of both the subject and object in activity. Here Heidegger gives the famous example of a hammering. A hammer is "ready-to-hand" because in its use, both the nail and the hammer sort of disappear and we are only left with the hammering. *Vorhandenheit* is a type of detached viewing of an object bereft of any phenomenological context. This is the mode of encounter for the scientist investigating an object under the guise of neutrality. However, Heidegger's point is that ultimately this is part of a metaphysics of presence and inherently problematic because of the prioritizing of a false subjectivity. Neither *Vorhandenheit*, nor *Zuhandenheit* seem to apply to Rollin's zoo animal analogy "there as our tools for amusement, not in their own teleological magnificence."[23] Granted there are difficulties with translation and others have translated these terms differently, but phenomenology requires precision. As a matter of fact one could even potentially say that it is the teleological focus on action that makes the objects "transparent" *Zuhandenheit*, ready-to-hand. For example, a burger could be ready-to-hand in the act of eating. The problem with Rollins focus on *telos* or goals of things, an Aristotelian instead of a phenomenological concept, is who determines the *telos*? A richer phenomenological explanation actually could have furthered Rollin's analysis. I don't want this to sound like a harsh critique, the language of phenomenology can be extremely difficult and technical and many surveys don't do it justice.

For my reader interested in the human-animal encounter in the zoo, I hope you will find a thoughtful examination of some important structures of this experience.

Chapter 1 will provide a crash course in phenomenology. It will put the method of phenomenology "on display." In this chapter, I will introduce the method of phenomenology that will be used to examine the phenomenological context of the zoo. I provide a history of the movement with a grounding in its origins from its founder Edmund Husserl and his transcendental method.

Chapter 2, "Phenomenology and the Life-World of Animals" will begin the conversation of major issues and advantages involved in applying phenomenology to the context of animals and people. In particular, the phenomenological position will be read with and against the work of the philosopher Thomas Nagel. Also, phenomenological empathy will be read alongside the contemporary work of the animal behaviorist Temple Grandin.

Chapter 3 will provide visitor descriptions and narratives of zoo visits and the visitors' motivations. This chapter will survey some common assertions and assumptions about zoos that need to be bracketed with a phenomenological analysis. Here the common beliefs about the zoo visit will be uncovered and made apparent. This is a "natural world" analysis, so it begins the phenomenological method by providing a stepping point from which to begin the analysis.

"Bracketing: A Trip to the Zoo," is the fourth chapter and is a short narrative of a zoo visit. This chapter utilizes the famous phenomenological reduction to call into question our assumption about what goes on in the zoo visit. Here I will tell the story of two famous zoo animals from my childhood and begin the process of a descriptive and praxis-oriented phenomenology.

Chapters 5 and 6 are explicitly communicative phenomenological explications. Here is where the real discoveries of praxis phenomenology are made. Chapter 5 will look at a rhetorical analysis of zoos through the trope of synecdoche. Chapter 6 will investigate the concept of play and empathy in the zoo setting. The focus of chapter 5, "Rhetoric and Synecdoche," is on the spoken word and, in particular, looks at the notion of naming animals in the zoo. Chapter 6, "Playing at the Zoo and Kinesthesia," examines the nonverbal communication taking place in the zoo.

The next chapter will begin a genetic phenomenology of the zoo investigating the troubling origins of the phenomenon and the changing cultural meaning of zoos around the world. A generative phenomenology will be conducted as well in this chapter, where there will be an investigation of the scientific notion of biophilia alongside the phenomenological concept of the *Lebenswelt*, or life-world.

Finally, there will be an epilogue, which will attempt to bring this all together to show how this phenomenological journey has answered the questions of why we go to the zoo asked in this introduction. The epilogue will provide a few recommendations for both zoos, and I also will point to some future directions and research as well here.

The purpose of this book is to provide a phenomenological analysis of zoo visits. The zoo is a controversial space that deserves a philosophical analysis without prejudgments and assumptions that it is extraneous at best, or even immoral. I ask the reader to put aside your own prejudgments you may have about zoos as you read this book. This book tries to look at what goes on in the experience of the zoo visit. I hope to explore the meaning of underexplored experiences I have observed in zoos. Looking at these experiences in the context of the history of zoos should allow us to make some phenomenological observations about the zoo visit. After engaging in these methodological reflections, I hope we will be able to answer the question: Why do we go to the zoo?

NOTES

1. Association of Zoos and Aquariums, "Public benefits of AZA." http://www.aza.org/public-benefits/.
2. Robert Mullan, and Garry Marvin, *Zoo culture*, (Chicago: University of Illinois Press, 1999), xii. www.waza.org/en/site/about-waza/who-we-are.
3. Kate Connolly, "Rejected at birth, Knut becomes Berlin zoo's bear essentials," The Guardian , March 23, 2007. http://www.guardian.co.uk/world/2007/mar/24/animalwelfare.germany.
4. Annie Leibovitz, "Leo and the bear," *Vanity Fair*, April 2, 2007. www.vanityfair.com/politics/features/2007/05/knut_slideshow200705.
5. "Polar bear controversy," reported by Richard Schlesinger, March 22, 2007. www.cbsnews.com/video/watch/?id=2599094n.
6. Tom de Castella, "Knut: A polar bear story," BBC News , March 22, 2011. http://www.bbc.co.uk/news/magazine-12805534.
7. David Crossland. "Farewell Knut, thanks for the memories," Der Spiegel , sec. Obituary, March 21, 2011. http://www.spiegel.de/international/germany/obituary-farewell-knut-thanks-for-the-memories-a-752247.html.
8. Ibid.
9. Ros Clubb, and Georgia Mason, "Animal welfare: Captivity effects on wide-ranging carnivores," *Nature* 425 (2003): 473–74, doi: 10.1038/425473a
10. Georgia Mason, and Jeffery Rushen, *Stereotypic animal behavior: Fundamentals and applications to welfare* (Cambridge, MA: Cabi, 2006).
11. Mary Beth Warner, "Polar bear will be given home at natural history museum," Der Spiegel , March 25, 2011. www.spiegel.de/international/germany/remembering-knut-polar-bear-will-be-given-home-at-natural-history-museum-a-753142.html.
12. Olin E. Myers, Carol D. Saunders, and Erik Garrett, "What do children think animals need? Aesthetic and psycho-social conceptions," *Environmental Education Research* 9, no. 2 (2003): 305–25, doi: 10.1080/13504620303461
13. However its roots are in the ancient discipline of rhetoric, making it as old as philosophy. But in the academy, rhetoric was largely ignored as a discipline and is one of the newest majors to have been widely taught in various universities.
14. For example, see Robert Cox, "The die is cast: Topical and ontological dimensions of the locus of the irreparable," *Quarterly Journal of Speech* 68 (1982): 227–39, doi: 0.1080/00335638209383609. Or, Thomas Farrell, and Thomas Goodnight, "Accidental rhetoric: The root metaphors of Three Mile Island," *Communication Monographs* 48: 271–300, doi: 10.1080/03637758109376063. Or, Christine Oravec, "Conservationism vs. preservationism: the 'public interest' in the Hetch Hetchy controversy," *Quarterly Journal of Speech* 70: 444–58, doi: 10.1080/00335638409383709.
15. Robert Cox, *Environmental communication and the public sphere* (Los Angeles, CA: Sage Publication, 2009).
16. Craig Waddell, *Landmark essays: On rhetoric and the environment* (Mahwah, NJ: Lawrence Erlbaum and Associates, 2007).
17. Peter Singer, *Animal liberation: A new ethics for our treatment of animals* (New York: New York Review and Random House, 1975).
18. Tom Regan, *The case for animal rights* (Los Angeles: University of California Press, 1983).
19. Stephen Spotte, *Zoos in postmodernism: Signs and simulation* (Madison, NJ: Fairleigh Dickinson University Press, 2006).
20. Bernard Rollin, "Through a frame darkly: A phenomenological critique of zoos," in *Metamorphoses of the zoo: Animal encounter after Noah*, ed. Ralph R. Acampora (Boulder, CO: Lexington Books, 2010), 57–66, 59.
21. Ibid., 62 ed.
22. Martin Heidegger, *Being and time*, trans. John Macquarrie and Edward Robinson (New York: Harpers, 2008).
23. Rollin, "Through a frame," 62.

Chapter One

Displaying the Phenomenological Method

It was 1932. Raymond Aron had just returned to Paris after spending some time studying at the French Institute in Berlin. He had gone to study phenomenology and, in particular, the works of Edmund Husserl to help with his dissertation in history. Soon after returning, Aron found himself in the café Bec de Gaz in the Rue Montparnasse with Jean-Paul Sartre and Simone de Beauvoir. The threesome would meet regularly and discuss a wide range of philosophical and political issues. The specialty of the house was an apricot cocktail. After ordering their drinks, Aron pointed to the glass and said, "You see, my dear fellow, if you are a phenomenologist, you can talk about this cocktail and make philosophy of it!" The idea that tangible objects in the world could be described through a philosophical process excited Sartre so much that he turned white in the face, leapt up, and went to Germany to begin a close study of phenomenology.[1]

I remember hearing this story as an undergraduate studying philosophy. It was the first day of my Existentialism class and was my initial exposure to the world of phenomenology. I remember having a reaction similar to Sartre's. I was fascinated by this acumen of phenomenologists. My friends often accused me, a student of philosophy, of wanting to construct castles in the sky, yet the phenomenological method was a method for investigating the bricks of the real world. I knew that this was the technique to discuss what really matters, and vowed to begin a close reading.

In this chapter I hope to provide a basic understanding of phenomenology necessary to understand the discoveries of the rest of the text. I hope to answer the following questions: What, generally speaking, is phenomenology? What specifically are the steps of the phenomenological method? Why use phenomenology for this specific study? And finally, what are the limitations of using phenomenology for this current study?

WHAT, GENERALLY SPEAKING, IS PHENOMENOLOGY?

Background

In 1927 Edmund Husserl, considered the father of phenomenology, was asked to write an article defining this branch of philosophical science for the *Encyclopedia Britannica*. Husserl ended up writing five different drafts and some fifteen pages to try to make phenomenology accessible to the layperson. Husserl was originally trained as a mathematician. This may account for why his writings often have the characteristic of some advanced theoretical calculus whose message takes some degree of familiarity and training to understand. I feel sorry for the layperson who came across Husserl's lengthy description of phenomenology in the *Encyclopedia Britannica* and who hoped to understand this philosophical system without already having been trained in philosophy. Only an especially gifted philosophical mind could have grasped it outright.

In order to begin my task of giving a simple description of phenomenology, it is important to begin with a caveat. Phenomenology can very broadly be described as a philosophy "whose point of departure has been immediate experience and whose method has involved an analysis of experience."[2] As Quentin Lauer points out, such a broad umbrella would include every philosopher from Aristotle through Descartes, and Augustine through Hegel. Therefore, it is my belief, like Lauer's, that any attempt to define phenomenology must begin with Husserl's conceptualization of his philosophical system. Since Husserl there have been many phenomenologists who have rejected some of his tenets and modified others. However, in order to render a definition of phenomenology, it is necessary to begin with a description of Husserlian phenomenology as phenomenology. It is my belief that a definition of Husserlian phenomenology for the layperson can be found in three central propositions: 1. "Consciousness is always consciousness of something." 2. "Phenomenology is a method of rigorous science." 3. "Phenomenology is an eidetic science." I believe that these three tenets are the key points to understanding the following statement by Husserl:

> This phenomenology, like the more inclusive pure phenomenology of experiences in general, has, as its exclusive concern, experiences intuitively seizable and analyzable in the pure generality of their essence, not experiences empirically perceived and treated as real facts, as experiences of human or animal experients in the phenomenal world that we posit as an empirical fact. This phenomenology must bring to pure expression, must describe in terms of their

> essential concepts and their governing formulae of essence, the essences which directly make themselves known in intuition, and the connections which have their roots purely in such essences. Each such statement of essence is an *a priori* statement in the highest sense of the word.[3]

"Consciousness is Always Consciousness of Something"

This statement brings us to the apricot cocktail of Aron, Sartre, and de Beauvoir as they sat in their 1930s café in Paris. Phenomenology's central concern is experience. Aron, Sartre, and de Beauvoir had a sensual experience of the apricot cocktail. They only knew it through their seeing it, tasting it, smelling it, touching it, and hearing of the glass itself as they clink them together in a toast. Prior to Husserl, philosophy imported preconceived notions into what we could know about the experience of the object of the cocktail. A popular epistemology was Locke's representational account of knowledge. According to Lockean representationalism (which Husserl attacks for being a form of psychologism), knowledge held that we had an inner copy in our mind of the object that existed outside and separate from the mind. Phenomenology rejects this separation of an object outside of the mind and completely separate internal copy in the psychic realm. Phenomenology is fundamentally antidualistic. The experience of the object cannot be stripped of our experiencing of said object. What this means is that the only way we can truly discover anything meaningful about the apricot cocktail is through our various sensuous perceptions of it, i.e. our seeing it, tasting it. As Husserl says:

> But experience is not an opening through which a world, existing prior to all experience, shines into a room of consciousness; it is not a mere taking of something alien to consciousness into consciousness. . . . Experience is the performance in which for me, the experiencer, experienced being "is there," and is there *as what* it is, with the whole content and the mode of being that experience itself, by the performance going on in its intentionality, attributes to it.[4]

So in referring to experience as "a consciousness always conscious of something," this implies a "performance" where the experiencer and the content of the experience are inseparable. "Consciousness" in this statement refers to the subject and his/her intentional gaze.[5] That "something" is the content of consciousness. Since this "something" is inseparable from consciousness it entails the content of all the various modes of consciousness itself. Therefore, this content is the correlate of that which is revealed through consciousness tasting, seeing, and so forth—the something which is tasted, seen, and so forth Since the modes of consciousness extend beyond sensual experience of the five senses to aspects such as remembering, projecting, imagining, and so forth, this "something" is also the remembered, projected and imagined and therefore that which is present to consciousness and not necessarily present in the physical world in the common use of the term "actually present."

"Phenomenology Is a Method of Rigorous Science"

Now in this broad description of the content of consciousness and the phenomenon of experience, it would appear that Husserl leaves himself open to the charge of abstract internalism or at least some form of idealism. An examination of Husserl's call of phenomenology to be a rigorous science should show this charge to be unfounded. In his Inaugural Freiburg Lecture Husserl calls for a close methodological examination of phenomena. "Pure phenomenology claims to be the science of pure phenomena. This concept of the phenomenon, which was developed under various names as early as the eighteenth century without being clarified, is what we shall have to deal with first of all."[6] This taking the phenomena seriously, as such, leads to a "science of consciousness."[7]

In his development of phenomenology, Husserl was combating dangerous trends he saw in psychology and other human sciences. Husserl constantly warns against the dangers of "psychologism." According to Husserl, psychologism was "the positivistic reduction of the idea of science to mere factual science."[8] The problem this reductionistic trend caused was a "'crisis' of science as the loss of its meaning for life."[9] Psychologism was equated with positivistic trends, which had been spread from the hard sciences to the human sciences. And this positivism "decapitates philosophy."[10]

So, far from wanting to develop some idealistic, abstract philosophical enterprise, Husserl saw his project as saving science by laying bare its true foundations and subsequently restoring its human meaning. To do this, Husserl uses the example of Galileo, "a discovering and concealing genius."[11] The great discovery and concealment of Galileo was his "mathematization of nature."[12] In order to conduct his modern physics Galileo treated the natural world in an idealized manner. Galileo wanted a systematic mathematization that would allow for the commensurability of nature.[13] In order to do this, nature was treated as a geometric perfection and idealized, so that it was separated of its subjective character. However, in the real world we do not really encounter perfect circles, squares, or numbers. The things of the world are always experienced as this thing here and subsequently never separate from the human subject. Actual perfect forms don't exist, therefore any measurement will always be an approximation. This was the discovery and concealment of Galileo—the mathematization of nature.

This mathematization is just taken for granted by the sciences and Husserl wanted to understand the sciences' foundation in order to better understand science itself. The problem was a substitution of a method for the actual world, which caused us to lose the meaning of the actual world, and thus created the "crisis of the sciences" for Husserl. So, Husserl wanted to save the sciences by making them rigorous and apodictic through phenomenology. By apodictic this meant that he wanted to go back to their foundations and understand their originary meaning. To do this, phenomenology uses a method of "bracketing" called the *epochē* in which the ontological statuses of all beliefs are voluntarily put aside in order for the phenomenologist to understand the meaning of said beliefs and, consequently, their impact on the scientific investigation itself. Therefore, phenomenology is a presuppositionless science.

There is no negative connotation of doubt with the phenomenological reduction, or *epochē*. What *epochē* literally means is a "leading back."[14] So phenomenology is a method of removing the sedimentations that have veiled experience. In order for this to be done, all ontological commitments must be suspended and then a detailed description of experience can take place. The naïveté of the lived everyday world is bracketed and the rich phenomenological world is revealed. The phenomenologist does this through a detailed methodology of description.

"Phenomenology is an Eidetic Science"

Ultimately phenomenology is concerned with more than "mere appearances." So, what is described in the phenomenological method? What the phenomenological method describes are the *eidea* or essences. Now these essences are not Plato's otherworldly essences. Instead the Husserlian essences are found in the flow of everyday experience through the phenomenological method. However, the constitution of these essences is within the horizon of the experience of the objects themselves. It is important to note that Husserl is not positing a dualistic split from otherworldly noumena. Instead eidetic science is the "field of assertions about essences, viz., of what is generally the case as given in pure 'seeing.' That is to say at first undifferentiated from the individually given universal objects."[15] It is in this grounded manner that the essences are given in an *a priori* manner. They are *a priori* in that they are prior to reflection.

Eidetic science is engaged in through a second *epochē*—an eidetic reduction. There is a bracketing off of everything in order to engage the "pure mental processes." This allows for an analysis that moves beyond the mere givenness of the factually occurring phenomena to the pure structures that make it possible to perceive such phenomena.

> For instance, the phenomenology of perception of bodies will not be (simply) a report on the factually occurring perceptions or those to be expected; rather it

> will be the presentation of invariant structural systems without which perception of a body and a synthetically concordant multiplicity of perceptions of one and the same body as such would be unthinkable.[16]

The meaning of this eidetic discovery is that eidetic science must be rational. So rationality is what comports the scientific aspect of eidetic science. But its anchoring in experience means it is intuitive as well.

This pure phenomenology, which gives us the *eidos*, is to be the foundation for philosophy because of its "all embracing" grounded methodology. So, phenomenology is simultaneously both "first philosophy" and a methodology in and of itself. Phenomenology is simultaneously both praxis and philosophy and therefore, not abstract or ideal. This all-encompassing laying bare of foundations is also what makes Husserl refer to in eidetic science as a "genetic phenomenology." Genetic phenomenology is equivalent with phenomenology as eidetic science. The central concern is the *a priori* revealing of the universal, flowing *eidos*.

Genetic phenomenology is extremely relevant to the current study of child development. This is because ontogenetic development is explained in the eidetic science. Husserl's phenomenological method may be the first philosophical method to explain the temporal change of meaning in our experience of objects. For example, the meaning of a pet is different for the child of six from what it is for the same person grown up at age sixty. Different experiences, histories, cultural shifts in meanings have led to this change. However, genetic phenomenology contains its own reduction that peels away the various sedimentations that have led to this change. Perhaps it is this genetic phenomenological method that makes Husserl utilize many examples involving children in his writings.

The role constitution plays in Husserlian genetic phenomenology is what necessitates that this investigation be grounded in a Husserlian phenomenology. The child's relationship with animals must be conceptualized in a non-static manner that permits an openness and flow of development. Husserl's genetic phenomenology's attention to the temporal allows for a stripping away of various sedimentations in order to reach the *eidos*. Since the experience of the child is of a child who is "growing up," I believe that Husserlian (meaning genetic) phenomenology must be the starting point.

WHAT SPECIFICALLY ARE THE STEPS OF THE PHENOMENOLOGICAL METHOD?

Natural World and *Epochē*

One of the peculiarities in the history of phenomenology is that a lot of Husserl's critics seem to have misunderstood him at the very start. Husserl's call that the phenomenological reduction be absolute created quite a controversy that caused many phenomenologists after Husserl to claim that this was impossible. This is the critique that Merleau-Ponty makes in the preface of his *Phenomenology of Perception.*[17] However, the phenomenological reduction or *epochē* is not the starting point of Husserlian phenomenology.[18] Husserl's phenomenology is concerned with a being in becoming. So in order to clear up any possible mistakes that may be made by taking up another style of phenomenology, let us begin our detailed examination of Husserlian phenomenology.

Husserl begins with an *empirical ego* engaged in the *natural world* with a consciousness in the *natural attitude*. The natural attitude is the everyday flowing prereflective experience that consciousness has. The natural attitude is full of presuppositions formed by the flow of our everyday life. In the natural attitude, consciousness takes for granted things-in-themselves, held to exist independent of consciousness. We "experience" objects without ever questioning the validity of their existence. The empirical ego is the ego of everyday consciousness engaged in a pretheoretical, prereflective, and, consequently, prephenomenological world. Phenomenology begins with a free choice by this empirical ego in the flow of the everyday natural world to engage in the *phenomenological reduction* or *epochē*.

Husserl's phenomenological method begins with a modification of Descartes's notion of universal doubt called the phenomenological reduction, or *epochē*. The *epochē* is different from the Cartesian notion of universal doubt. What the *epochē* does is "bracket off" the world so that we can examine the primordial essences. Instead of doubting the existence of any particular phenomenon, we simply hold off from reaching a judgment. We put the phenomenon in-itself aside and abstain from making any conclusions about its existence—that is, its existence independent of subjectivity.

This setting aside of the world is a radically presuppositionless attitude which enables us to examine how consciousness spontaneously encounters the world. In the phenomenological *epochē*, when I set aside my judgments about the existence of a house, for example, I do not lose any of the properties inherent in the essence of a house. The house itself, whether real or imagined, is in no way changed by the *epochē*. The *epochē* instead directs my reflective regard to how my consciousness perceives and deals with this house as phenomenon.

The *epochē* is the first step in founding the apodictic certainty of the phenomenological sciences. It requires a "leap of faith" on the part of the subject engaged in the natural world. There must be a willingness to look at the world in a new way and a subsequent commitment to search for foundations without presuppositions in the move from the natural attitude to the phenomenological attitude. The *epochē* requires the phenomenologist to step outside of the natural attitude and be willing to examine experience in a totally new light. As Husserl says when describing his "first" methodological principle: "Genuine science, must neither make nor go on accepting any judgment as scientific *that I have not derived from evidence*, from 'experiences' in which the affairs and affair complexes in question are present to me as 'they themselves.'"[19] The *epochē* asks us to put aside this world which is immediately present to us. Again it is extremely important that it be distinguished from the Cartesian method of doubt.

Husserl's relation to Descartes is also extremely important. There is an explicit rejection of Cartesian dualism in phenomenology. While it may be possible for the empirical ego in the natural attitude to experience a quasi-split of mind and body, in the phenomenological attitude the ego experiences the world in its lived totality. "I, this One among Others, experience the world as a surrounding world."[20] We are always at the center of the world and have our own horizons from which we deal with the phenomena of our experience, but the world still is intersubjective. Therefore, phenomenology precludes both the possibility of either a detached materialism concerned merely with the *res extensa*, or a detached idealism concerned merely with the *res cogita.* "Husserl himself actually claimed that the existence of the object was phenomenologically irrelevant (Hua 19/59, 358, 387, 672)"[21] But also for Husserl phenomenology is not merely pure introspection. Yet, the objects in the world cannot be discussed independently of their disclosure to a subject.[22]

Phenomenological Constitution

Let us examine an example that reveals the important role that *perception* plays in phenomenology. If you have ever driven on a United States interstate with a child who lives in an urban area you will no doubt receive an alert as you pass the first dairy farm. "Look, a cow," can become a frequent refrain. Let us begin a phenomenological description of the conditions for the possibility of the child's perception of the cow.

The child's view of the cow, or any phenomenon in the world, is limited by the facticity of the child's existence. The child's position in relation to the actual object, the cow, is in a sense a limitation. Yet, the temporal experience of consciousness in the phenomenological attitude is not limited. Experience flows and allows the child to apperceive even that which cannot be seen. Inevitably a young child will moo after alerting you to the cow. Because of part of the child's past experience with cows mooing and future expectations, the child will intuit that cows moo. Mooing is part of the essence of a cow for a child. The ease of this auditory connection of animals with sounds is one of the reasons that the sounds of animals are one of the first lessons taught to children in everything from songs like "Old McDonald," to "spin and say toys" and games that parents play with their children.[23] These past and future experiences of a phenomenon that constantly flow into my conscious perception of that phenomenon are known by the phenomenological terms of *retention* and *protention*.

The common phenomenological example of the *constitution* of a house will help make these terms more clear. If I stand in the front of a house, I am unable to simultaneously perceive the back; however consciousness does not question whether there is a back. Instead, because of my past experience with houses, and my future experience, I intuitively know that it is part of the essence of a house to have four sides, a floor, and a roof. The anticipations, protentions, are rooted not only in rules of perception, but also come from culture. Furthermore, the rules of perception and apperception will be important in establishing an intersubjective grounding of the life-world because the structures of apperception and perception have allowed us, through experience, to establish meaning—including that which we cannot actually see, but what we intuit to be so.

Husserl credits Franz Brentano with the discovery of *intentionality*. The structure of intentional consciousness occupies a central position in the entire Husserlian opus and thus Brentano was a major influence on Husserl. Through phenomenological analogues of perception, we move beyond the mundane passivity of the natural attitude and discover *intentional consciousness*. I become aware that in perceiving it is I who constitute, and always have constituted, meaning. The phenomena have been and always will be part of my consciousness. The phenomena have meaning only in so far as they are constituted by consciousness. When I perceive a book in the phenomenological attitude, I bracket off its very ontological status, so that what I am left with is the meaning of the object as constituted by consciousness. When I encounter the book I do not first ask myself, or in any way concern myself with, whether or not the book exists. Instead, I constitute the meaning of the book. It is this particular book, which happens to be my favorite, so I read it. Or it is some book I found on the street, which I don't care about and so use it as a doorstop. As my consciousness constitutes the meaning of the book for me, I at no time concern myself with the question of whether or not it actually exists independent of me.

We still need a tighter clarification of the concept of constitution here. Constitution is the "process" by which the perceived comes into perception. Constitution has a dyadic meaning of passive synthesis and active performance. Those who wish to interpret Husserl as mere idealist emphasize the creative performance of constitution by subjectivity. However, this ignores the role facticity plays in constitution. The subject does not "make" the object *ex nihilo*; instead the object is passively given to the subject in the mundane world. The constitution of the cow, or book, by the child is necessitated by an encounter with this cow in a mode of consciousness given to the empirical ego. Yet, after the phenomenological reduction, it is apparent that this given passivity is inseparable from the constituting subject. And as the structure of *Ideas Pertaining to a Pure Phenomenology and to a Phenomenological Philosophy* , shows, constituting subject, world, and intersubjectivity[24] are irrevocably intertwined. So, subject, world and community are all implicated in Husserlian constitution.

Husserl had a bit of the revealer and concealer in him when it came to the concept of constitution. While the discovery of constitution is one of the major achievements of the phenomenological method, it also is not very clearly defined by Husserl himself. In fact, it appears that he has at least two different types of constitution. The two types are *passive* or *static constitution* and *genetic* or *phenomenological constitution.*[25]

In order to keep the two types of constitution clear in my mind I use the analogy of seeing something with two or three dimensions. Static constitution is kind of like a square. It looks at the givenness of an object to experience and the manner it is given to the ego. Genetic constitution adds a third dimension of time and therefore is more like a cube.

The key to understanding static constitution is the two key phenomenological correlates of *noema* and *noesis*. Noema and noesis are the unified duality that is the structure of intentionality. Noema is the "objective pole" of intentionality while noesis is the "subjective pole." That the laws of intentionality are grounded in the noema is not to be understood in a Kantian phenomenalist way as an object that exists, in itself out there; as a matter of fact, it does not even have to be an external object.[26] The noemata of perception vary depending upon the observer's relation to the object, memory of it, or even imagination of it. The noema is that which is constituted by consciousness. As a result and in a manner of speaking, the noemata are outside of the act. "The noema is distinct from the act in the sense that it does not really exist within consciousness, as the act does."[27]

If the noemata exist in the "plane of sense perception," then their noetic correlates exist in the plane of action. The noetic is the act of constituting the noemata and establishing its meaning for consciousness. The noetic is centered in the ego that constitutes meaning as such. However, the ego as encountered via the noesis has changed from the empirical ego that is simply living in the experience of the natural attitude. The noetic finds a new ego that no longer resides in the natural attitude because it is now aware of the active power of constituting intentionality, and now simultaneously both resides and does not reside in the world.

This noematic and noetic axis are the essential tools of static or passive constitution. Now what is genetic or phenomenological constitution? The supplement to *Ideas Pertaining to a Pure Phenomenology and to a Phenomenological Philosophy: Third Book* contains an essay that Husserl meant to be a transition between the second and third book. At the end of this essay Husserl is obviously writing against Heidegger when he explicitly says, "*on-*

tology is not phenomenology."[28] In this piece, the crux of Husserl's attack on Heidegger is his failure to incorporate genetic, or phenomenological constitution into his ontology. Genetic constitution adds a third axis to constitution in that it considers constitution in its temporal flow.

> The ontological mode of consideration is so to speak katastematic. It takes the unities in their identity, and for the sake of their identity, as something fixed. The phenomenological-constitutive consideration takes the unity in the flow, namely as unity of a constitutive flow; it follows up the movements, the flows, in which such unity and every component, side, real property of such unity is the correlate of identity. This consideration is in certain measure kinetic or "genetic": a "genesis" that belongs to a totally different "transcendental" world than does the natural and natural scientific genesis.[29]

So, the critique we see here of ontology, and Heidegger, is the failure to recognize genetic or phenomenological constitution. There is a controversial move here that Husserl makes. Genetic constitution is transcendental. It is a constitution that goes beyond the natural world and takes unities in their temporal flow. This phenomenological analysis is also where Husserl writes of kinesthesis, or the movement of bodies. Ontology's constitution is purely descriptive and immanent. It is stuck in the natural world like a Newtonian apple. Genetic constitution, with its transcendental nature, is historical and therefore requires a viewing of the object and its various modalities in history.

For Husserl phenomenology, if it is to be phenomenology, must attend to origins. We are born into a world of meaning and our constitution of the world matures as we do. This is why Husserl illustrates many of his points with examples of children and infants to show how meanings come into being. The transcendental move for Husserl is a necessary one, for there is a "residue" left over after the transcendental reduction in that we have always been part of an ongoing conversation from our birth. My constitution of the world has always been influenced by the meanings made by those around me.

Heidegger, on the other hand, in his entire corpus, never writes on children or development. His Dasein points to a mature and fully formed being. And his explicit critique of Husserl's notion of the transcendental ego is that it is not Dasein. Now as many people know, *Being and Time* was never finished and Heidegger intended to concentrate more on time and being in the unwritten part, so it is possible that his critique of Husserl's transcendental phenomenology may have softened if he had indeed written more on time in a phenomenological sense as he intended. In his lectures and letters to associates like Karl Lowith and Karl Jaspers, Heidegger expresses an open hostility to Husserl's transcendental phenomenology, even calling it a "sham philosophy."[30] For Heidegger the transcendental ego is a break with the concrete descriptive method of the phenomenological tradition, because to him it is a "theoretical" abstraction that arises from some "triviality" that Husserl wants to pursue. However, Heidegger misread the meaning of the transcendental ego in part due to his failure to pay attention to maturation. If one wants a closer read of the conflict between Husserl and Heidegger, I recommend the edited volume *Psychological and Transcendental Phenomenology and the Confrontation with Heidegger (1927–1931)*, which contains a lot of primary material. For an explication of particular elements of the conflict, I recommend the work of Burt Hopkins or Steve Crowell.[31] In particular, I have to agree with Hopkins that Heidegger seems to have missed "the nuance"[32] in understanding the differences in Husserl's notion of psychic and transcendental immanence. In the next section, I will attempt to give more "nuance" to Husserl's transcendental phenomenology.

Transcendental Phenomenology

Husserl considers the transcendental move to be the most important for phenomenology. Transcendental phenomenology requires a whole new reduction—a transcendental one. This reduction is performed by the *transcendental ego*. It is a strange sort of reduction because it is an *epochē* on one's self, or own ego. A transcendental reduction is a move from the specificity of the empirical ego in the natural world to a pure consciousness. "The transcendental reduction neutralizes the 'ontic residue' of a consciousness conceived psychologically, namely with a real stream of consciousness. '*Epochē* allows me a universal reflection on my self-being.'"[33]

The reflection upon my constituting of meaning causes me to both lose myself and recover myself. I must step beyond the empirical ego of the natural attitude and become a monadic, concrete, and continuous ego, which the act of constitution requires. "The ego is himself existent for himself in continuous evidence; thus, in himself, he is continuously constituting himself as existing."[34] Likewise, I must step beyond myself as pure self and lay claim to the universal character of this transcendental ego.

The origin of the transcendental reduction is such that it appears to be in danger of drifting into a form of transcendental solipsism because Husserl's phenomenological discovery begins as Descartes did—with the self. However, Husserl's transcendental ego is not a mere transformed substance. The transcendental reduction, when followed through to its end, finds a reflective structure that moves beyond the facticity of individual existence and through the use of empathy into the realm of a universal intersubjectivity. As Husserl says in the *Cartesian Meditations*:

> I, with my life, remain untouched in my existential status, regardless of whether or not the world exists and regardless of what my eventual decision con-

> cerning its being or nonbeing might be. This Ego, with his Ego-life, who necessarily remains for me, by virtue of such *epochē*, is not a piece of the world; and if he says "I exist, *ego cogito*," that no longer signifies, "I, this man, exist."[35]

The transcendental ego exhausts itself in its individuation and must obtain its eidetic structure through a universal idealism. By overcoming the transcendental solipsism that is initially contained in the transcendental ego we achieve a new transcendental intersubjectivity.

The process by which transcendental solipsism gives way to transcendental intersubjectivity begins with the recognition that the Other as an alter ego is immediately experienced differently than the phenomena of objects that are also given to consciousness. The transcendental *epochē* does not allow for the presupposition of the Other, and therefore in our quest to determine the eidetic structure of the ego, Husserl discovered that "[i]t should be noted that in the transition from my ego to an ego in general, neither the reality nor the possibility of an inclusion of Others is presupposed. In this case the scope of the *eidos* ego is determined by a self-variation of my ego. I imagine myself as if I were someone else. I do not imagine Others."[36] The transcendental ego utilizes empathy to go beyond oneself and not only constitute the alter ego as in my consciousness, but ultimately sees that the alter ego is a constituting body as well. From this transcendental empathy I can now understand the alter ego as a unified essence like me, or a human being.

The transcendental reduction begins with an empirical factual ego. The individuality of this ego is what is transcended when it is bracketed from the lived-world through reflection that allows for the discovery of the ego's eidetic structure. In the transcendental reduction, the eidetic aspect of the transcendental ego and its universal aspect go beyond the individual ego. Finally, the eidetic aspect can become ours. This is another example of how the transcendental ego is a "recovering in me of what is *not* me."

The final phase of the transcendental reduction when followed through is the reduction to my "ownness sphere." This is sometimes referred to as the "primordial reduction."[37] The main question in this phase now becomes, "What is the transcendental meaning of the alien subject?"[38] As transcendental ego we are concerned with the constitution of the alter ego.

The Other, through the phenomenological method, has been bracketed up to this point. I experience the world as lived through my body. I experience the Other through her body, that I immediately perceive as similar to mine. Transcendental empathy sets the alter ego free. Through analogical apperception originally established in our perception of noematic objects, I can claim to be able to put myself into the alter ego's place and vice versa. The here and there of myself in relation to the alter ego become interchangeable. The social context of existence allows the alter ego to reveal herself through her embodied acts. Each smile, tear, or laugh I perceive of the alter ego is like looking through the windows in a house inside which I have never been. Eventually if I look through the windows enough and peer into all the rooms, I can make a good claim to know what the house on the inside actually looks like—without ever setting foot inside of it.[39]

The transcendental ego, through analogical apperception, first understands the alter ego as a physical being in the same way that I am in the world. The analogizing continues on into the psychic realm where transcendental empathy allows for the interchangeability of the alter ego with me despite our different positions in space and time. The next step is to return to constituting the world from this recognition of the Other. The meaning of the Other then transcends itself as the monadic unity human being. The final step is to constitute the "we" and return anew to the natural world and reexamine the world once again from the perspective of transcendental intersubjectivity. The result is an establishing of the *life-world* where the constitutional projects of humanity are examined in the context of their historical and cultural realities, which have now received apodictic certainty when the transcendental intersubjective notion of humanity is applied to the natural and primordial worlds.

Life-World

So what exactly is this life-world, or *Lebenswelt*? Perhaps John Wild, for whom the concept was central for his philosophy, put it best when he said, "This word cannot be exactly and objectively defined."[40] Eugene Fink inserted a sentence that would help explain this difficulty into the opening section on the life-world in *The Crisis of European Sciences and Transcendental Phenomenology*, "There has never been a scientific inquiry into the way in which the life-world constantly functions as subsoil, into how its manifold prelogical validities act as grounds for the logical ones, for theoretical truths."[41] The best place where Husserl performs a close inquiry into the nature of the life-world is in the third section of *The Crisis of European Sciences and Transcendental Phenomenology*.

Husserl begins this section by chastising Kant for not having discovered the life-world. According to Husserl, Kant's problem was that he did not examine his suppositions coming from the traditions in which he built his theory. The most succinct critique comes in the following note in this section:

> Yet this (fault) does not lie in (a psychological) beginning. (In fact) the first thing Kant (should have done, if he) had taken the everyday world as the world of human consciousness, was to pass through psychology—but a psychology which allowed the subjective experiences of world-consciousness actually to come to expression as they showed themselves experientially.[42]

In looking at the presuppositions of Kantian philosophy and objective science, Husserl makes his discovery of the life-world. Prior to the Kantian philosophy and an objective science, "the world is always in advance."[43] The life-world is the grounding from which any scientific endeavors take place. So, the life-world is a return to the natural world that had been bracketed by the *epochē*, after the transcendental move of phenomenology. Husserl feels the purity of the universal ego allows for a proper examination of the life-world, not as the impossibility of a detached observer in objective science, but from both within and outside this world (transcendent). The first discovery is that the transcendental ego makes about the life-world is that the life-world has "a *general structure*. This general structure, to which everything that exists relatively is bound, is not itself relative."[44] Through the transcendental reduction, Husserl returns to the relative natural world in his life-world investigation and discovers that it has a universal structure.

Let us follow Husserl in pointing out some of the main features of the life-world. The first of these structures is that the world is "spatio-temporal." Prior to any objective science, there is already a spatio-temporal structure, which allows for any sort of investigation.

The second structure of the life-world is that it is bodily. Now "bodily" here does not mean bodily in the way of Galileo's physics of bodies. Instead, Husserl claims "the bodies familiar to us in the life-world are actual bodies."[45] For Husserl, Galileo's physics of bodies had been guilty of abstraction through the mathematization of nature. The bodies of the life-world are "real bodies" prior to physics, and thus Husserl's transcendental move into life-world analysis is a realism. However, as bodily, the life-world allows for an investigation of motility of a special type—kinesthesis. Husserl's understanding of kinesthesis simply is the study of the experience of living moving bodies in the life-world.

Just as the intentional consciousness after the first *epochē* had a correlative structure of noema and noesis, so does the life-world. There is a "formal structure" of the life-world that has the correlates of "thing and world on the one-side and thing-consciousness on the other."[46] This structure begins like the structure of intentional consciousness in making the point that "the world is the universe of things."[47] However, prior to the things being given to the world, there already always exists universally the condition for these things existing—the life-world. So, by examining the presuppositions about the life-world in the transcendental *epochē*, Husserl makes the important distinction in noting the key difference between consciousness of things and the world. The world is the universal grounding condition for the particular things to be grasped by consciousness, and consequently the world is singular.

> Every object has its possible varying modes of being valid, the modalizations of ontic certainty. The world, on the other hand, does not exist as *an* entity, as

> an object, but exists with such uniqueness that the plural makes no sense when applied to it. Every plural, and every singular drawn from it, presupposes the world-horizon. This difference between the manner of being of an object in the world and that of the world itself obviously prescribes fundamentally different correlative types of consciousness for them.[48]

So it just wouldn't make sense to talk about plural worlds in the phenomenological sense of world as originary ground.

Another fundamental feature of the life-world is that it is teleological. This discovery is made by Husserl in implementing a focus on "how" the life-world and its objects are given to consciousness. Husserl's discovery here mirrors Kierkegaard's emphasis on the "how" of indirect communication of the subject as opposed to the "blasphemy" of the what of direct communication. Husserl notes that in living in the world, we are always "directed" towards this world in all our actions. As the universal field in which all our actions are grounded, the world provides both "coherence" and the horizon of all our goals. In discovering a *telos* structure in the life-world, Husserl simultaneously gives an explanatory concept for life itself inside of the life-world. Husserl explains this thusly, "Among the objects of the life-world we also find human beings, with all their human action and concern, works and suffering, living in common in the world-horizon in their particular social interrelations and knowing themselves to be such."[49]

Kierkegaard is apropos here, because Husserl's method of transcendental *epochē* reveals a paradox of the life-world. That the life-world, as grounding horizon for all the objects in it, allows for the human subject to be studied qua objects. Yet, the human subject also transcends the life-world in order to investigate it and transforms it with a primordial act symbolic of all acts. This paradox is resolved for Husserl in transcendental intersubjectivity. The life-world is always the meaningful ground for all humanity.

> What was lacking was the phenomenon of the change of signification of (the form) "I"—just as I am saying "I" right now—into "other I's," into "all of us," we who are many "I's," and among whom I am but *one* "I." What was lacking, then, was the problem of the constitution of intersubjectivity—this "all of us."[50]

In the *epochē*, each person was treated as an object of investigation, in the life-world, yet as "human being" in the life-world, each person is also her own "I" with her own "accomplishments." So the final and most important character of the life-world is that it is the ground of communalization. The life-world is social. However, this sociality is not taken for granted, but foundational in the very center of our investigation into the meaning of the life-world as such. The apodictic certainty with which Husserl describes the life-world is what separates his life-world investigation from other later phenomenologists. It is the greatest achievement of Husserlian phenomenology, and this is why Husserl felt transcendental phenomenology was the most important achievement for phenomenology.

SUMMARY AND ISSUES OF PHENOMENOLOGICAL METHOD

The purpose of this previous section was to introduce the terms needed to understand how Husserl has made phenomenology first a grounding for the intersubjective and the explication of this transcendental intersubjective humanity in the life-world.[51] This last section has been presented as it is because of my belief as to how Husserl should be read. There is a theory asserting that Husserl's philosophical works have distinct phases, which if they do not conflict, at least represent a focus on completely different notions. The other theory of interpretation of Husserl's works is that there is a unity that the later concept of life-world provides, and as a matter of fact, is needed to truly make sense of phenomenology.[52] In *World and Life-World*, Balazs Mezei claims that both J. N. Mohanty and he are followers of the unity method of interpretation.[53] Gerd Brand and Roman Ingarden on the other hand represent the modification theory. My reading of Husserl sides with Mohanty and the unitary theory of Husserlian phenomenology.

The brilliance of Husserl is in the rigor with which he pursues his methodology. In order to properly understand Husserl and phenomenology, you must take a transcendental approach and view phenomenology as a unity. The strength of phenomenology is in its following the systematic approach to the end. To stop short of the grounding of transcendental intersubjectivity in the life-world is to miss not only phenomenology, but the existence that it properly describes.

NOTES

1. Paraphrased from Lester Embree, "The natural-scientific constitutive phenomenological psychology of humans in the earliest Sartre," *Research in Phenomenology* 9, no. 11 (1981): 41. See also: Simone de Beauvoir, *The prime of life*, (Chicago, IL: Paragon, 1992), 112.

2. Quentin Lauer, "Questioning the phenomenologists," *Journal of Philosophy* 58, no. 21 (1961): 633.

3. Dermot Moran, *Introduction to phenomenology* (New York: Routledge, 2000), 1–2. As quoted from: Edmund Husserl, *Logical investigations, Vol. 2* (New York: Humanities Press, 1970), 249.

4. Moran, *Introduction to phenomenology*, 6. As quoted from: Husserl, *Formal and transcendental logic*, 233–34.

5. For the phenomenologist, intentional has a specific technical meaning that will be explained later.

6. Edmund Husserl, "Pure phenomenology, its method and its field of investigation," presented at Inaugural Freiburg Lecture, 1917. www3.baylor.edu/~Scott_Moore/essays/Husserl.html. See also: Peter McCormick, and Elliston Frederick eds., Husserl: Shorter works , trans.

Robert W. Jordan and intro. H. L. Van Breda (Notre Dame, IN: University of Notre Dame Press, 1981), 9–17.

7. Ibid., 16.

8. Edmund Husserl, *The crisis of European sciences and transcendental phenomenology* (Evanston, IL: Northwestern University Press, 1997), 5.

9. Ibid.

10. Ibid., 9.

11. Ibid., 52.

12. Ibid., 23.

13. Commensurability of nature refers to the ability or an attempt to measure nature.

14. Michael Theunissen, *The other: Studies in the social ontology of Husserl, Heidegger, Sartre, and Buber* (Cambridge, MA: MIT Press, 1984), 24.

15. Edmund Husserl, *Ideas pertaining to a pure phenomenology and to a phenomenological philosophy: First book*, (Boston, MA: Kluwer Academic Publishers, 1989), 6.

16. Edmund Husserl, *Psychological and Transcendental Phenomenology and the Confrontation with Heidegger (1927–1931)* (Boston: Kluwer Academic Publisher, 1997), 165.

17. Maurice Merleau-Ponty, *Phenomenology of perception* (New York: Routledge, 1999).

18. In my first detailed Husserlian investigation in my Master's thesis I made this same mistake of beginning with *epochē*. I am indebted to the great phenomenologist, Dr. Marilyn Nissim-Sabat for her reminder that Husserl starts in the natural world.

19. Edmund Husserl, *Cartesian meditations*, trans. Dorion Cairns (The Hague: Springer Press, 1967), 13.

20. Husserl, *Ideas: First book*. See also: Theunissen, *The other*, 27.

21. Dan Zahavi, *Husserl's phenomenology* (Stanford, CA: Stanford University Press, 2003), 61.

22. Some critics have claimed that Husserl's *epochē* and his search for foundations led to an "internalism." Hubert L. Dreyfus, David W. Smith, Ronald McIntyre, and Dagfinn Follesdal have reached this conclusion by interpreting the noemata as "mere representations" and consequently split from the world. Like Zahavi, I feel that this conclusion is based on a misinterpretation of the noema and consequently reject this "California school, or West Coast interpretation of Husserl." See Zahavi, *Husserl's phenomenology*, 58.

23. This auditory example should help to clarify the phenomenological conception of perception. In phenomenology, perception is not merely the visual (as a matter of fact it is not merely the sensual). Instead, perception is the presentation of the object in its manifold forms to consciousness.

24. Empathy plays a key mediating role between subject, world, and Others. For example, see Section 2, chapter 4 "The Constitution of psychic reality in empathy," in Edmund Husserl, *Ideas pertaining to a pure phenomenology and to a phenomenological philosophy: Second book* (Boston, MA: Kluwer Academic Publishers, 1989), 170–81. Moran uses the terms static and genetic

25. Moran, *Introduction to phenomenology*, 166–68.

26. Aron Gurwitsch, *Studies in phenomenology and psychology* (Evanston, IL: Northwestern University Press, 1966), 131–34. See also: Robert Soloman, *Phenomenology and existentialism*, (New York: University Press of America, 1980), 231–35.

27. Ibid., 133.

28. Edmund Husserl, *Ideas pertaining to a pure phenomenology and to a phenomenological philosophy: Third book* (Boston, MA: Nijhoff, 1980), 117, §6.

29. Ibid.

30. Edmund Husserl, *Psychological and transcendental phenomenology and the confrontation with Heidegger (1927–1931)*, trans and ed. Thomas Sheehan and Richard Palmer (Dordrecht: Kluwer Academic Publishers, 1997).

31. Burt Hopkins, "The Husserl-Heidegger confrontation and the essential possibility of phenomenology: Edmund Husserl, psychological and transcendental phenomenology and the confrontation with Heidegger," *Husserl Studies* 17, no. 2 (2001): 125–48, doi: 10.1023/A:1010771906591. See also: Galt Crowell, "Does the Husserl/Heidegger feud rest on a mis-

take? An essay on psychological and transcendental phenomenology," *Husserl Studies* 18, no. 2 (2002): 123–40, doi: 10.1023/A:1015540519310.

32. Hopkins, "The Husserl-Heidegger confrontation," 137.

33. Victor Velarde-Mayol, *On Husserl* (Belmont, CA: Wadsworth, 2000), 15.

34. Husserl, *Cartesian meditations*, 66.

35. Ibid., 25.

36. Ibid., 72. See also Theunissen, *The other*, 40–1.

37. Theunissen, *The other*, 55–6.

38. Husserl, *Cartesian meditations*, 94.

39. It is important that this is a metaphor and not meant to be taken literally. Phenomenology would not make objective claims on the Other. We can never experience what the Other experiences *qua* Other.

40. John Wild, *Human freedom and social order: An essay in Christian philosophy* (Durham, NC: Duke University Press, 1959), 65. See also: William McBride, "John Wild and the life-world," in *Analecta Husserliana, vol. XXVI*, ed. Eugene Kaelin and Calvin O. Shrag (Secaucus, NJ: Kluwer Academic Publishers, 1989), 103.

41. Edmund Husserl, *The crisis of European sciences and transcendental phenomenology* (Evanston, IL: Northwestern University Press, 1997), 124.

42. Ibid., 117.

43. Ibid., 110.

44. Ibid., 139.

45. Ibid.

46. Ibid., 142, § 37.

47. Ibid.

48. Ibid., 143, § 37.

49. Ibid., 146, § 38.

50. *Ibid.*, 182, § 54.

51. Velarde-Mayol, *On Husserl*, 81.

52. Balazs M. Mezei, *World and life-world: aspects of the philosophy of Edmund Husserl* (New York: European University Studies Press, 1995), 8–9.

53. Mohanty confirmed this unity theory in both a presentation and conversation at the Sixth Annual Phenomenology Roundtable held at Temple University June 29, 2006.

Chapter Two

Phenomenology and the Life-World of Animals

Now that we have an outline of the steps of the phenomenological method we can look at whether it is the proper method for investigation in this study. First of all, what is being studied? This book examines the experience of humans and has a particular focus on children visiting the zoo and why we keep going back? During their development, a significant number of children form very strong bonds with animals. These relationships are revealed through the unique ways that children connect with animals. Up to this point they have escaped a close philosophic scrutiny and a close examination of those in the field of communication who analyze a variety of relationships. A child's relationship with animals is often overlooked or simply dismissed as being unimportant. What are the advantages of a phenomenological investigation of this subject? These advantages can be revealed by looking at the human-animal bond very broadly with an eye on the notions of presuppositionlessness, genetic phenomenology, and life-world investigations.

PRESUPPOSITIONLESSNESS

Presupposition—Aristotelian Hierarchies

Why have philosophers and those who study communication tended to overlook the experiences of children and animals? Since the days of Aristotle there has been a tendency to define "man as the rational animal." By exclusion, women, children, and animals have been considered outside the domain of rationality. Therefore, since they are outside of reason, they are outside of the domain of rational inquiry itself. That is why historically children and animals may have gone without any rigorous philosophic investigation.

Aristotle is an apropos starting point. For it was Aristotle, who was one of the earliest and closest observers of nature. His observations and writings on hyena sexuality were accepted as fact until very recently.[1] He wrote vociferously on nature and animals—not only in his works specifically on nature—but even had an eye turned to nature in his *Rhetoric* when he remarked about the rhetorical significance of "beast fables" for making points in public address because of the ease in which they can be constructed and because of the widespread familiarity the audience will have with the subjects.[2] Aristotle coined the terms: oviparous (from egg), viviparous (from uterus), and vermiparous (from worm), which are still used today. As George Gaylord Simpson noted in his *The Principles of Taxonomy*, "Yet Aristotle's philosophy is still with us today; it is the backcloth to our attitudes to animals, and, in fact, to the whole way we live and think."[3]

Aristotle's rigorous study of nature would have been desired by the phenomenologist. However, he imported a presupposition of a hierarchy into his *Nature*. In this hierarchy, animals were viewed as existing for our use. So, in his detailed descriptions of animals and their habitats, Aristotle was looking for their essence or cause with a final *telos* of being for a human purpose already in mind. The presuppositionless phenomenological attitude, would have to have a different starting point. We would begin by noting that in the everyday natural attitude of many (particularly Western) societies, animals are viewed as existing purely for human use. Yet, within the first reduction this presupposition, would need to be bracketed in order to determine the essences at the root of the human-animal experience. This study, as phenomenological, must jettison this hierarchy in the examination of animals.

While animals have been included in the discourse of Aristotle (but in a subservient position to humans), children have been flat out ignored in much philosophical discourse. I feel this absence has been in large part due to the exclusion of feminine voices from philosophy. Because the first place where children really get rigorous philosophic consideration is in connection with the notions of "motherhood" and feminity in the works of philosophers such as Arendt and de Beauvoir. If to be human means to have reason—then children have philosophically been viewed as not having fully developed reason and therefore, being outside the realm of philosophical engagement. Philosophical chauvinism identifies the "ideal" human being with the human most full of reason and historically that domain has been falsely ascribed as the domain of the white, adult male. This study begins with a value claim that the experiences of children are worthy of discussion for their own merit. This may sound like a commonsense claim, but viewed from within the natural attitude of the tradition of philosophy, this was a highly controversial claim.

Presupposition—Nagel and Radical Separation

Another problem with philosophical discussion of animals is conclusions about a radical difference between the experience of animals and humans. We need look no further than the father of modern philosophy, Descartes, to see such a view. Descartes was guilty of claiming that animals were merely automatons, without souls, and unable to feel pain. Fortunately this radical claim is no longer given much merit, yet the radical divide between animal and human experience exists. Let us closely examine one place where this divide is clear—Thomas Nagel's famous, "What is it like to be a bat?" essay.[4]

In this essay, Nagel's main purpose is to counter claims made in the philosophy of mind by reductionist physicalists who attempt to solve the mind-body problem by reducing all mental phenomena to a physical explanation. According to Nagel, "there is no reason to suppose that a reduction which seems plausible when no attempt is made to account for consciousness can be extended to include consciousness. Without some idea, therefore, of what the subjective character of experience is, we cannot know what is required of physicalist theory."[5]

Nagel then goes on to propose that what is missing in the physicalist theory is an accounting of the subjective nature of consciousness, what he very broadly connects with the notion of "what it is like to *be* something." The example of a bat is then given to show what a physicalist theory of mind is missing. The bat is chosen both because of its mammalian closeness to humans and its alien behavior and ways of orienting to the world—specifically a bat's echolocation, the sonar that allows the bats to perceive the world as these animals navigate through the air.

> But bat sonar, though clearly a form of perception, is not similar in its operation to any sense that we possess, and there is no reason to suppose that it is subjectively like anything we can experience or imagine. This appears to create difficulties for the notion of what it is like to be a bat. We must consider whether any method will permit us to extrapolate to the inner life of the bat from our own case, and if not, what alternative methods there may be for understanding the notion.[6]

For Nagel we can never get to the point where we could actually understand what it is like for a bat to perceive the world qua bat. At best we can imagine what it would be like for us to be a bat, but that is something different from bat experience as the bat. Nagel feels this points to a bigger issue—the difference between facts and conceptual schemes.

> My realism about the subjective domain in all its forms implies a belief in the existence of facts beyond the reach of human concepts. Certainly it is possible

> for a human being to believe that there are facts which humans never *will* possess the requisite concepts to represent or comprehend. . . . But one might also believe that there are facts which *could* not ever be represented or comprehended by human beings, even if the species lasted forever—simply because our structure does not permit us to operate with concepts of the requisite type. . . . Reflection on what it is like to be a bat seems to lead us, therefore, to the conclusion that there are facts that do not consist in the truth of propositions expressible in a human language. We can be compelled to recognize the existence of such facts without being able to state or comprehend them.[7]

From his conclusion of the subjective experience of bats and their radical incompatibility with the human perspective, Nagel comes to the conclusion that there "obviously" are facts that exist independent of human experience. Nagel goes on to conclude that the "mysterious" difference of these facts (between bats and humans) does not offer a refutation of physicalism per se, but instead leave us with these facts that are unexplainable in physicalist terms because they do not take into account the different phenomenological apparatus in which different subjects perceive the world.

Nagel concludes with a call for an "objective phenomenology."

> Setting aside temporarily the relation between the mind and the brain, we can pursue a more objective understanding of the mental in its own right. At present we are completely unequipped to think about the subjective character of experience without relying on the imagination—without taking up the point of view of the experiential subject. This should be regarded as a challenge to form new concepts and devise a new method—an objective phenomenology not dependent on empathy or the imagination. Though presumably it would not capture everything, its goal would be to describe, at least in part, the subjective character of experiences in a form comprehensible to beings incapable of having those experiences.[8]

Nagel's call is for a phenomenology that sets aside empathy and imagination and can arrive at "the facts" of subjective experience in an objective way.

Nagel's essay has had a major impact on writings on the philosophy of mind that are out of bounds of the present study. Yet, despite his radical phenomenological claims, he has not been taken up critically by most phenomenologists. Nagel's attack on empathy, imagination, and notion of fact have direct bearing on the phenomenology engaged in over the course of this book project, and his phenomenological missteps need to be made clear before the phenomenological meaning of the zoo visit can be explicated.

Nagel is right to criticize the physicalist account of mind for not being able adequately to describe mental aspects of experience. For the phenomenologist, the dualistic mind-body problem is a presupposition that falls beneath the *epochē*. We do not have split experiences and thus a complete split of mind and body is impossible.[9] However, Nagel's insistence on the objective existence of facts outside of the human subject is problematic. Phenomenology asks how can we know anything except through consciousness? That is not to say that phenomenology does not leave open the possibility of new experiences. There is always the possibility of new experiences, but it is meaningless to talk about facts existing independently of the human experience, for it is only through the possibility of human experience that they receive any meaning.

To make this clearer, let us utilize a Zen meditation puzzle that focuses on nature. A beginning Zen Buddhist may be asked to meditate on "whether if a tree falls in a forest and no one is around, does it make a sound." For Nagel the answer seems to be yes. There is a realism of "facts" beyond the subjective domain. However, a Husserlian phenomenologist would approach this problem differently. In fact, I can't say whether the tree makes any sound. That is not to say that it does or doesn't make a sound, but the meaning of the fact of a sound a tree makes in falling in the forest is only conceivable through the subjective correlates of a subject that possess the capacity for hearing. We cannot make any claims of factual objects existing independently of subjective experience; however, this does not say that they do not "actually" exist, but just show that experience must be structured in a unitary manner and cannot be falsely split into neat and separate spheres of body and mind.

Nagel's bifurcation of experience is clear in his statement, "But one might also believe that there are facts which *could* not ever be represented or comprehended by human beings, even if the species lasted for ever—simply because our structure does not permit us to operate with concepts of the requisite type." He is right, that we shouldn't doubt that there are facts that exist outside of the representation or comprehension of human experience, but that does not mean that they do actually exist and should be treated as such. For the structure of "facts" for a phenomenologist is such that they could not be meaningful without human representation or comprehension.

So the problem is that Nagel is subscribing to a physicalist notion of facts, while phenomenology requires an entirely different notion. In this study of the child-animal bond the notion of fact cannot be reduced to mere physical qualities. This is the dangerous move made by the psychologists engaged in psychologism and the communication scholar engaged in abstract quantification. The *epochē* prevents the phenomenologist from importing preconceived notions of a physical/mental split to the human and phenomena/noumena to experience. While it is possible for noumena to exist it is impossible for them to be understood, represented, and so forth, without a human subject. Therefore, it is impossible to reduce the child-animal bond to mere physicalist behaviors, because the subjective life-world is always the grounding condition in which experience happens. That is why the essential structures of the human animal bond cannot be known through a behaviorist lens that falsely treats behavior as independent fact separate from consciousness. The possibility of a separation of mental from physical facts is a presupposition that is accounted for in order to get closer to the "reality" of actual experience. This project is an eidetic science, not a factual one. And as such it is concerned with the real essences of experience and not with the abstractions of physicalist facts.

PRESUPPOSITION—NAGEL'S ATTACK ON IMAGINATION AND EMPATHY

Imagination

Now there is a second problem in Nagel's essay with his attack on imagination and empathy. Nagel's phenomenological misstep can be accounted for by a close reading of Husserl's notion of empathy in *Ideas, Second Book.*[10] Nagel criticizes the physicalists for their failure to explain what is meant by "are" in the statement "mental states *are* states of the body; mental events *are* physical events."[11] Yet Nagel does not fully explain what is meant by "be" in his question "what is it like to *be* a bat?" To be a consciousness with experience means to grasp the things of the world in their manifold appearance. The manifold ways that things appear to consciousness include the various actual sensual modes of perception, but also the imagination. As Husserl says of the phenomenological task of constitution,

> The task is rather to examine, *in eidetic intuition, the essence of the experienced in general and as such*, precisely as it is made explicit in any experience, whether carried out actually or imaginatively . . . in order then to grasp intuitively, in the unfolding of the intentions essentially involved in such an experience, the sense of the experienced as such—the sense of the relevant class of regional objectivities—and to express this sense in rigorous analysis and description.[12]

Children have very active imaginations. Their lives involve many rich fantasy experiences. This is particularly revealed in imaginative play, which will be discussed later. The phenomenological notion of "reality" is different from "actuality." Therefore, the imaginative graspings of children's consciousness are a "reality" of consciousness even though these imaginative graspings may not be "actual." Imaginative constitutions are real ways that things unfold to consciousness and therefore are within bounds of the phenomenological direction of this study. This may seem counterintuitive, but it is an important phenomenological starting point. Imaginative consciousness is a "real" manifold experience of consciousness. Traditional folk psychology, psychologism, and naturalism would dismiss this imaginative role and place it out of the bounds of investigation. For the phenomenologist who does not begin with such presuppositions, the imaginative graspings of children have an actual impact on the child's consciousness despite their not having an "actual" factual content. I like to use the example of the frog and leaf from Gestalt psychology to help clarify this point. When I am out walking along with my wife and think I see a frog up ahead I will stop and say, "Oh look, a frog," or "Wow, a frog," or "Be careful not to scare the frog." We may intentionally move around the frog in a broad sweep in order to avoid scaring him. However, when we approach the frog, we see that the frog is "actually" a leaf. Now in the moment that our imaginations take the leaf as frog there are "real" experiences of consciousness that cause us to react in a certain way talking about the frog, moving in a sweep in order not to disturb the frog, and so forth (Or if I was back in my younger days—maybe trying to catch the frog). Conscious experience and its actions cannot be completely understood with merely the "actual" leaf thing. Phenomenological understanding of consciousness requires an accounting of the "real" imagined frog.

Empathy

So, we see that a thorough understanding of consciousness in an "objective" and "real" sense requires imagination. However, it also requires empathy. According to Husserl there are two worlds implicated in spiritual constitution: the "surrounding world or external world of the communal spirit" and the "merely subjective sphere of the solitary subject."[13] How does Husserl arrive at the notion of these two worlds? One of the most intriguing sections of Husserlian phenomenology concerning empathy is Husserl's section on "The Constitution of Psychic Reality in Empathy" in *Ideas Pertaining to a Pure Phenomenology and to a Phenomenological Philosophy: Second Book.* Husserl begins this section on constitution by describing the appearance of those things that are given to consciousness in space and time. Other people are an interesting given thing because they are given as "two-fold unities,"[14] both a thing and a psychic individual.

> Given *eo ipso* with the act of apperceiving something as human is also the possibility of mutual relations, communication, between (hu)man and (hu)man. Then also the identity of nature for all men and animals. Given furthermore are the more simple and more complex social connections, friendships, marriages, unions; these are connections instituted between men (on the lowest level, already between animals.)[15]

The notion of empathy is often misunderstood. It is mistakenly regarded as mysterious, when it actually has a distinct and important phenomenological grounding. That is why Nagel wants to jettison empathy in his "speculative" call for an "objective phenomenology." However, an "objective phenomenology" requires a move from natural science to human science in order to grasp the natural world in its entirety.

An "objective phenomenology" would necessarily begin with an object's *primal presence.* Husserl defines primal presence as "*original perception* to the lived experience in which a subject has the perceived object given in original presence, then that means the object is there 'actually,' 'in the original,' and is not merely 'co-presented.'"[16] What does it mean that this object is there "actually?" The actual primal presence has two aspects. First of all, the perceptual object is potentially primally present for everybody. "The totality of objects that can be given as primally present and that make up a domain of common primal presence for all communicating subjects is *nature in the first and original sense.*"[17] An objective phenomenology of perceptual objects given in this manner would lead properly to the natural sciences. However, the other aspect of primal presence is the subjective. As Husserl says, "Each subject belongs here itself along with all its acts, states, noematic correlates, and, furthermore, with the corporeality and the properties, or faculties, constituted in it in the inner attitude."[18] Subjectivity is primally present in correlation with the objective givenness to all of an object. A properly objective phenomenology cannot exclude an investigation of subjectivity. What we have here is the origin of the human sciences.

Husserl maneuvers between these two correlates of primal presence with an analysis of corporeality and the body. In primal presence, bodies are given to me. However, I am also a body. Being an enfleshed, corporeal body means that I have "the experience of localization." This localization is the body's co-present givenness through senses and psychic acts. I am a subject who has sensual and mental experiences through my body. This is all given in primal presence.

An objective phenomenology would be incomplete in primal presence because the Other is also primally presented as body. Part of having a body means having a localization in the world. "Animalia"[19] are given in a unique way, they are primally presented as objects with corporeal bodies, but they have a localized here like mine—an interiority. Thus an objective phenomenology of the constitution of animalia must take into consideration *appresence* given to all living bodies like mine. Through the unity of soul[20] and body, the psychic is also localized in space and time and a proper area of investigation for objective phenomenology. Appresence is the shifting of my localized "here" to the Other's "here" in which everything else in the world serves as a background of "there." This analogizing appresence is the origin of empathy, and with this starting point we have the condition for a proper objective phenomenology—a phenomenology of the human science that treats the human as unity of subjective and objective, and follows primal presence through to appresence.

Husserl gives an example of an objective phenomenology of a cat playing that shows the possibility of Nagel's call for an objective phenomenology void of imagination and empathy.

> I see a playing cat and I regard it now as something of nature, just as is done in zoology. I see it as a physical organism but also as a sensing and animated Body, i.e., I see it precisely as a cat. I "see" it in the general sense ordinarily meant when speaking of seeing. . . . In its own way, the cat also is seen, and, in the seeing, its existence as this animal, cat, is experienced. This experience has the kind of imperfections corresponding to the fundamental character of experience of animals. But, as always, the cat is present there in the flesh—specifically, as a physical thing with sensing surfaces, sense organs, and so forth. The stratum of sensation is not there as something beside the physical thing; what is there is a Body, a Body which has physical and aesthesiological qualities as one. Likewise, the Body is also experienced as Body of a soul, and the word "soul" indicates again a founded stratum of qualities, and of course one that is still higher. The soul is not there as extended over the Body in the manner of being "localized" in the proper sense; . . . the psychic is . . . something at the Body or in it, lacking only distinguishable separate location. . . . The excess of reality beyond the mere physical thing is not something that can be separated off by itself, not something juxtaposed, but something *in* the physical thing; thus it moves "along with" the thing and acquires its spatial determination by its being in something which is itself spatial.[21]

What Husserl shows in this example is how an "objective phenomenology" of the cat must be done of the cat qua cat. What that means is that the objective phenomenology must include a phenomenological analysis of both primal presence and appresence. The cat as body is a physical body that moves. However, as moving body of animalia, the cat possesses a psychic quality which moves *in* the physical body. As embodied, this psychic holds a localized space and time, and thus has an objective quality for objective analysis. This is why we have objective evidence to talk of the cat as an acting agent just like me. The inability to perceive the cat as having a psychic state like me shows a failure of social development and a failure of the faculty responsible for appresence explaining the connection between animal abuse and other forms of abuse conducted toward humans.

Nagel is right to critique the physicalists for leaving something out in their explanation of experience—the mental or psychic. Just as the physicalists held the presupposition that the mental could be reduced to physical, so does Nagel hold a presupposition in his speculative proposal that unitary experience of mental and physical must be split to engage in an "objective phenomenology" when we see that a rigorous and objective phenomenology must jettison the reductionist dualism that makes physicalism possible in the first place. When carried through this objective phenomenology reveals the appresence of Other animalia in the world like me.

GENETIC PHENOMENOLOGY

Today I saw a woman walking a yellow Labrador retriever across the street. The dog was remarkably old. The dog's slow, arthritic gait was an obvious testament to having lived a long, full life. The slow, hobbled shuffling of the dog was painful to watch and brought back vivid memories of the numerous canine companions I had lost to old age. Yet, despite the obvious difficulty the yellow Labrador had in walking, he was wagging his tail—revealing his predilection toward the walk.

The phrase "dog years" is a reminder of how much shorter the average canine lifespan is compared to humans. Yet, often a child in the family and the dog will come into the family at approximately the same time. Two main reasons the dog may be brought into the house are as surrogate/practice children, or *for* the children when the child is young. Child and dog often "grow up" together. The advanced speed of the dog's growing means that oftentimes the first place a child will encounter death is with the family pet. The encounter of old age and death is even more sped up in other companion species such as hamsters or fish.

At Brookfield Zoo's Hammill Family Play Zoo there is a Memorial Wall, where children can bring pictures of their deceased pets to put up. The children will draw the pictures and tell you stories such as "This is Rocky. He is a Boxer. He used to love to go on long walks and tug on a stick. Rocky got old and sick and wasn't able to play anymore. Now he's in doggie heaven and can go for walks and play again." Over the ten or so years that Rocky and the child were together, Rocky changed, yet was still the same dog. Certain methodological approaches to the child-animal bond would either focus on the dog as same or as different. For example, a behaviorist approach would look at the child's different responses to the various stimuli the dog provided. Therefore, the reduction to different behaviors would not necessitate a conceptualization of the dog as unitary whole. Other theories that smuggled in preconceptions of the animal or pet prior to the relationship would have problems relating to the changing states of the dog for the bond. A genetic phenomenology is important for this study because genetic phenomenology deals with the constitution of the object as it is lived in the stream of experience by the child. Therefore, we see this same dog in different phases throughout life. Genetic phenomenology would begin with the initial constitution of the object and allow for a mapping of new constitutions of the pet as body over time.

The dog is not the only one aging. The child is growing as well. Another advantage to genetic phenomenology is that it deals with temporal differences in constitution for the same subject over a lived history. The child is also both the same subject, yet different, over the course of time. The development of the child affects how the child constitutes her or his relationship with animals. Growth of the child needs to be considered across the history of the subject as well. Child development is not merely the physical growth and maturation of the child. It also is not merely the accumulation of new experiences by the child. Child development is an internal *growing with* the body. Development is a restructuring of the self-same subjectivity that allows for the potentiality of new horizons of experience.

The question of whether the subject's development is "innate" or "learned" is problematic for genetic phenomenology. It crosses the horns of this dilemma by saying it is both. For genetic phenomenology recognizes the child is born with certain potentialities and internal structural schemes. However, these structures undergo development through an outward learning process. This is very similar to how Piaget describes the development of intelligence.[22] The effect of experience is a sedimentation of the ego. Development is the residue of experience that attaches to the ego. Genetic phenomenology is a peeling away of the layers of the ego to understand the essential structures that make the possibility of said experience possible.

In the fourth Cartesian Meditation, Husserl focuses on the role development plays in constitution and points to genetic phenomenology as a solution.[23] Genetic phenomenology becomes of the utmost importance for understanding ego development in constitution and, consequently, child development. The originary grasping of earlier histories is part of our current constitution. Each earlier view is sedimented below the surface of our current constitution. Genetic constitution allows for the making clear of these earlier views and why Husserlian phenomenology is so friendly to considerations of child development. As Dermot Moran points out:

> Before we had an adult perception of things in the world we saw them as children. The child's way of experiencing these objects is somehow layered in with our adult way of seeing. Therefore, for Husserl, it is possible to grasp how an infant grasps objects, not by doing third-person psychology but by uncovering in oneself the layering of acts which show up the passive genetic acts which must have occurred. We cannot put ourselves back into the pure passivity of the infant's experiencing, nor do we have to rely on child psychology, for instance, but we can repeat these experiences in ourselves through the discovery of the eidetic laws which constitute them.[24]

Genetic phenomenology is an uncovering of originary *eidos* experiences in ourselves. Against the backdrop of the life-world, these essences are then able to be extended toward Others through analogical apperception. Thus we have a theory of child development without presuppositions that begins with my own "genetic acts."

LIFE-WORLD INVESTIGATION

> Now though we must (further) make evident the fact that if the life-world itself is a "structure," it is nevertheless not a "purposeful structure." —Edmund Husserl[25]

I am of the opinion that the life-world is the most important discovery of Husserl's phenomenological method. Remember that Husserl felt that there had never been a systematic inquiry into the nature of the life-world. Yet the life-world is that world that exists prior to all investigations. This study is not an investigation of the "scientific-world" but a cultural and historical investigation prior to the settling of any "purposefulness." If this were a scientific inquiry I would "merely" consult the tradition of child psychological research and animal behavior research and build upon the "scientific truths" of these disciplines to determine the meaning of zoo visits. Or perhaps conduct a survey of visitors via the quantitative analysis of much of visitor studies research. However, that would not be a life-world inquiry. I want to know what is the meaning of the relationship that causes us to return to the zoo in the context of the world as lived in the life-world.

Let's imagine a little five-year-old girl playing with the cat, as described (above) by Husserl. What are the structures of the life-world that we would need to consider in order to endow the life-world with its appropriate meaning?

Life-World as Bodily

The little girl and cat may play with a ball. The child may roll the ball across the floor and the cat will wait and then pounce on the ball. Jumping on the ball and hitting it with front paws simulating the way the cat would "toy" with a mouse. The cat will then retreat allowing the girl to get the ball and roll it again. The little girl will laugh and get the ball and begin the whole process again. This game will usually continue until the cat tires of the play, or the girl, unable to resist, tries to grab the cat for a hug, or initiates petting.

There is a sense that cat, girl, and ball exist in-the-world in a way that hinges on the importance of bodies in a motion.[26] According to Husserl, "everything that exhibits itself in the life-world as a concrete thing obviously has a bodily character, even if it is not a mere body, as, for example, an animal or a cultural object, i.e., even if it also has psychic or otherwise spiritual properties."[27] Without physical being and motion there would be no play. Cat, girl, and ball are all three bodies in the world. Yet, the cat and girl are bodily in a way that the ball is not. Husserl distinguished two different ways of being bodily in the world. The first was the manner that the ball was merely in the world, as a *Körper*. For Husserl, Körper bodies are bodies in "the geometric or physical sense."[28] Both the child and the cat are "present in the flesh."[29] To be "present in the flesh" means to be bodily in the world in a second way, the way of the *Leib*. Leib is being bodily as a "living body."[30] To be a living body means to have a personal subjective way of not merely being seen, touched, smelled, heard, but a body that sees, touches, smells, hears, and tastes. The Leib is an active body in an originary sense. A closer analysis of *Körper* and *Leib* will be done later in this work.[31]

Life-World as Empathic

The condition or possibility that the girl and cat can "play together" with the ball is made possible by the fact that by its very structure the life-world is necessarily empathetic through and through prior to thematic investigation. The play between girl and cat requires the girl and cat to see the objects of the world in the same way at that engaged moment. The ball must be a "play thing" for both cat and girl. If one of the participants does not see the ball, or it is an object of fear for one of the participants, then there can be no play. Not only must that ball be an object for play for the girl and the cat, but they must view the other as being engaged with the ball in the same way. So, the girl sees the ball as play thing and imagines the cat as viewing the ball as play thing; and likewise for the cat.

It may be perceived as a controversial statement to attribute empathy to the cat. But I mean it in only the most basic sense and can back it up. First of all, the play is not always initiated by the girl. The cat will initiate play by hitting the ball in front of, or toward the girl. In order to initiate play, there must be a recognition that this body, like me, is capable and willing to play. Also, the "language" of cats (as with dogs and most animals) is primarily bodily. How is it that the cat can predict the location of the ball? The cat must have a bodily attunement to the body of the ball and the action of the girl in order to so quickly read the motion and know where to pounce. The cat has a bodily awareness of the ball as body and of the girl as a body that not only moves like the ball, but is also a source of action that could grab the cat, pet the cat, or initiate some other action. The cat can read the body posture and determine the girl's intent. The cat can read the girl's body while she rolls the ball. If the girl rolls the ball, leans forward and reaches for the cat in order to squeeze her in a hug, the cat will not react to the ball, but will run away.

The life-world has a special structure of empathy. To navigate the life-world as Leib in our everyday life requires a firmament of empathy upon which the potential for interaction must take place. The insights of Temple Grandin come out of an intuited empathetic grasp of animals. Temple Grandin is one of the world's leading experts on animal behavior. She also has autism. In her book, *Animals in Translation*, Grandin claims that having autism gave her an extrasensitivity to see the world as animals see it. First, the animals helped her deal with her autism. As a teenager she describes visiting a ranch with her aunt and seeing cows placed in a "squeeze chute."[32] The squeeze chute was a big, metal V the cows would enter and it would then press against them, and the ranchers would then be able to give the cows their shots. Grandin was amazed at how calm the cows were. She realized that the pressing chute of the squeeze must be providing a soothing effect, like a massage. Since she was having trouble relaxing in school, she decided to make herself her own squeeze chute to help calm down.

> I got through my teenage years thanks to my squeeze machine and my horses. Animals kept me going. I spent every waking minute that I didn't have to be studying or going to school with those horses. . . . I wish more kids could ride horses today. People and animals are supposed to be together. We spent quite a long time evolving together, and we used to be partners. Now people are cut off from animals unless they have a dog or a cat.[33]

Grandin used this empathy to help both animals and herself. Her personal squeeze machine was her way of incorporating something that worked for the animals into her own life. Her professional practice of helping solve animal behavior problems on farms resulted from her ability to "think like the animals." She would get in the animals pens, runs, cages, etc; get down on the animals level; take pictures and translate for the farmer what might be bothering the animal. Here is how Grandin characterizes her advice:

> I always tell people: whenever you're having a problem with an animal, *try to see what the animal is seeing* and experience what the animal is experiencing. There are lots of things that can upset an animal—smells, changes in routine, exposure to things he hasn't experienced before—and you should consider all of them. Anything in the sensory realm can upset an animal.[34]

Phenomenologically speaking the life-world is the realm that makes it possible to see the world as the animal does.

Life-world as Intersubjective

The cat and girl do not play in a vacuum. There are historical and cultural achievements that have enabled the moment of this play to take place. Historically, the domestication of animals had to have already taken place and specifically the domestication of this animal called a cat. The ball they play with had to be created in a culture that placed importance on the play of people and pets. The child was raised in a specific family, a specific community, and a specific culture that passed along meanings of play, ball, cat, pet, and so forth. There are certain gendered role ascriptions for little girls and the type of play appropriate for girls, depending upon the culture in which the play is taking place. The girl and cat play in an intersubjective life-world community.

The status of children and animals in relation to the life-world community is problematic and one that Husserl considered. To what extent can we place children and animals in the life-world community when they are incapable of exercising the *epochē* themselves, or engaging in transcendental subjectivity?

> And what about children, even those who already have a certain amount of world-consciousness? After all, it is only from the mature and normal human beings who bring them up that they first become acquainted with the world in the full sense of the world-for-all, that is, the world of culture. And what about animals? There arise problems of intentional modifications through which we can and must attribute to all these conscious subjects.[35]

The role of children and animals in phenomenology was handled by Husserl, not in a closed manner that answered the actual transcendental status as either not having any transcendental subjectivity at all, or being able to objectively describe that transcendental objectivity from the outside. Instead Husserl said, "the point is not to secure objectivity but to understand it."[36] So, the life-world itself must become a philosophical object of inquiry for phenomenology. As a subject of inquiry, the point is not to secure "objective knowledge" that would treat the life-world as something different from what it is—living, changing and meaningful. For the life-world to truly be understood as life-world, the starting point of investigation must be a philosophic inquiry into all the various living communities of vibrant interaction that are the grounds of possibility for our transcendental subjectivity. Transcendental subjectivity and intersubjectivity ultimately constitute the life-world.

For the phenomenologist, the subject of inquiry is not some interiority, or idealism. Instead, the interest is in *how* we establish meaning in our various communities. So, in this phenomenological study, no claims are made to have "objectively secured" the knowledge of what is actually happening for the girl and the cat in their play. Instead, the knowledge of the structures and conditions of how their play takes place are described. With this shift in reason comes the move away from a singularity to intersubjectivity. The phenomenological investigation of the play between the girl and cat is never a singular horizon, but always already invokes manifold number of historical, actual, and potential communities. The intersubjective character of the life-world is *how* the life-world is to be understood.

Limitations and Controversies in Phenomenological Method

There are a couple of controversies that need to be mentioned in this closing stage about the methodology of this study. A thorough investigation of these controversies would each spawn several more books in their own right and are not the central focus of this text. The first and most obvious controversy is the use of Husserlian phenomenology as the phenomenological method used. Very simply, my reason for doing this is best summed up in the Mission Statement of the Phenomenology Roundtable:

> Husserl's phenomenology thus already presupposes many of the insights of his students. This is not to say that all knowledge is to be found in Husserl's writings alone. Rather, we recognize Husserlian phenomenology as our ground and as the proper method of philosophical analysis and work from this perspective in dialogue with other phenomenological perspectives.[37]

As a result of the methodological decision to utilize Husserlian phenomenology, two other controversies may automatically be attributed by critics of Husserl who have rejected the possibility of a total and complete *epochē* and transcendental subjectivity. It is my hope that the steps of the previous sections have shown my preference for these two methodological features as being fundamental aspects of the phenomenological method as such.

NOTES

1. Stephen Glickman, "The spotted hyena from Aristotle to the Lion King: Reputation is everything," in *Humans and other animals*, ed. Arien Mack (Columbus: Ohio State University Press, 1995), chap. 5.
2. Aristotle, *Rhetoric*, trans. George A. Kennedy (1991), 276–77.
3. George Gaylord Simpson, *Principles of animal taxonomy.* As quoted in Juliet Clutton-Brock, "Aristotle, the scale of nature, and modern attitudes to animals," in *Humans and other animals*, ed. Arien Mack (Columbus: Ohio State University Press, 1995), chap. 1.
4. Thomas Nagel, "What is it like to be a bat?" *The Philosophic Review* LXXXIII (1974): 435–50.
5. Ibid., 437.
6. Ibid., 438.
7. Ibid., 441.
8. Ibid., 449.
9. Despite analytic thought experiments actual experience involves both mind and body. Hence when attempting to describe total and complete mind they talk about "floating" spirit, which in itself contains a physical referent.
10. Specifically see: Edmund Husserl, *Ideas pertaining to a pure phenomenology and to a phenomenological philosophy: Second book* (Boston, MA: Kluwer Academic Publishers, 1989), 170–80.
11. Nagel, "What is it like?," 446.
12. Husserl, *Ideas: Second book*, 97.
13. Ibid., 208.
14. Ibid., 170.
15. Ibid., 170.
16. Ibid., 170–71.
17. Ibid., 171.
18. Ibid., 171.
19. Husserl's term for living animals, including humans. Ibid., 172.
20. Husserl uses this in a broad sense here, not in a specific religious manner.

21. Edmund Husserl, *Ideas: Second book*, 185–86.

22. There are many similarities between Piaget's genetic epistemology and genetic phenomenology. Piaget's emphasis on schemes, development, perception, critique of empiricism, revision of science, the acquisition of mathematical concepts, and the importance of time all have Husserlian points of overlap.

23. Edmund Husserl, *Cartesian meditations*, trans. Dorion Cairns (The Hague: Springer Press, 1967), 79, §38.

24. Dermot Moran, *Introduction to phenomenology* (New York: Routledge Press, 2000), 167.

25. Edmund Husserl, *The crisis of European sciences and transcendental phenomenology* (Evanston, IL: Northwestern University Press, 1997), 382. See Appendix 7.

26. Remember the earlier note that these motive bodies do not exist in the abstract way of Galileo's physics, but as actual bodies.

27. Husserl, *The crisis*, 106.

28. Ibid., 50. See footnote 15.

29. Husserl, *Ideas: Second book*, 185–86, §49 a.

30. Husserl, *The crisis*, 50. See footnote 15.

31. See chapter 4 on children and animals in zoos.

32. Temple Grandin, and Catherine Johnson, *Animals in translation* (Orlando, FL: Harcourt Inc, 2005), 4.

33. Ibid., 5.

34. Ibid., 23.

35. Husserl, *The crisis*, 187.

36. Ibid., 189.

37. Marilyn Nissim-Sabat and David Fryer, "Mission statement of the Phenomenology Roundtable," ratified by all the members through discussion, 2. Published as essay 38 in Cheung Chan-Fai, Ivan Chvatik, Ion Copoeru, Lester Embree, Julia Iribarne, and Hans Rainer Sepp, eds., "Essays in celebration of the founding of the Organization of Phenomenological Organizations," (2003), www.o-p-o.net.

Chapter Three

Let's Go to the Zoo

Natural World Description of Visitor Narratives

"What do you want to do today honey?"

"Let's go to the zoo?"

"Why do you want to go to the zoo?"

"Because I love the animals."

"How does going to the zoo let you 'love the animals'?"

"Because I can be with them."

I had the previous conversation with my three-year-old daughter last summer. We are zoo members and visit the zoo about five to ten times a year. We had not visited the zoo in a while, so I was curious about what prompted her zoo request. At age three, short-term memory rules much behavior, so a common occurrence about what to do the next day would be to repeat the previous day's experiences if they were half-way pleasurable. So, naturally I was surprised at the request for the zoo visit since we hadn't visited the zoo in several weeks.

My daughter's perceptiveness and linguistic skills are quite advanced for her age, but these comments shouldn't surprise many parents who probably have had similar conversations with their own children. While working for the Communication Research Department of Brookfield Zoo, I had a number of similar conversations with children (and even adults) who expressed similar motivations for wanting to visit the zoo that day. This chapter is concerned with sharing the stories of the many zoo visitors I have observed throughout the years.

THE NATURAL WORLD—"TO THE THINGS THEMSELVES!"

On Method

The method of the phenomenologist involves an immersion in the experience to be examined. Dive in as fully as possible. A problem many philosophers have is that they analyze and make judgments that prejudice their experience and block them off from fully engaging in the experience to be analyzed. There is a flow to the everyday experience in the natural world that requires a full participation in order to experience the phenomenon as lived. An overly analytic or critical mind can take the subject out of the flow of the experience and prevent a grasping of the phenomenon. Now that does not mean that the phenomenologist cannot be critical, or recognize inherent problems in particular experiences. It only means that in order to grasp a particular experiential phenomenon, the subject must begin in the flow of the natural world. This chapter will attempt to lay out some of the natural world beliefs about the zoo experience. This may seem like "common knowledge," but the point of the phenomenological method will be to make clear the everyday assumptions of the natural world. It is not to say that they will be overturned, or ruled incorrect, but that they must be made explicit in order to be bracketed.

One of the founding methodological principles of the ethnographic method as it has been taken up in anthropological practices is the idea of the "participant observer." The ethnographer will put aside personal and cultural biases and become immersed in the culture they are attempting to understand. There are close affinities between phenomenology and the social-cultural practice of ethnography.[1] The anthropologist engaged in ethnography must try to experience a cultural event via lived participation. It is the observations that come from these participations that the ethnographer attempts to describe. Acumen is needed to richly describe these experiences. It is a process that Clifford Geertz has characterized as "thick description."[2]

Let me give an example of a failure in "participant observation" that I had while teaching an introductory-level, intercultural communication course. The course was supposed to be service-learning, so students were doing work in low-income, predominately African-American neighborhoods as we read various primary text narratives of inner-city life. My students were invited into the community by a local community organizer who set up multiple youth activities to find healthy and supportive activities and creative outlets for children who had been abandoned by the community that had boarded up outlets and services for them to engage in.

One amazing program that our community partner started in Hazelwood was called the KRUNK program.[3] KRUNK stands for Kreating Realistic Urban New-School Knowledge and is an amazing hip-hop afterschool music and dance program. The program was created to give children a creative outlet to create and produce urban music as an alternative to the streets. These kids are really amazing and their professionalism has earned them invitations to play all over the country. My introductory intercultural students were given a crash course in ethnographic methods and told to observe one of the KRUNK performances as a participant observer and then write a "thick description" of the role music plays in the particular cultural community the students observed. The first time I went with students and everything went exactly as I hoped. The show was rocking, and my students participated, had a great time, and had wonderful interactions with the community members. However, the next semester I sent students by themselves to the show. Afterwards, the community partner talked to me and told me I needed to do something about my students. I asked him what happened. He told me that about ten of my students showed up for the show. They all sat together and watched the show with clipboards in hand, writing for the entire hour without participating. Now the closest way I can describe the KRUNK performance is akin to an African-American Baptist Church. In order to participate you need to move, jump, and sing. If you sit silently you aren't participating. My students were "detached observers," not participant observers, and thus they missed out on the true kinetic energy of the performance. And even worse, there clinical behavior negatively impacted the participatory character of the usual performance of the KRUNK group.

My reason for sharing the previous story is that to begin a phenomenological analysis you must jump into the experience. The natural attitude of the phenomenological method is prereflective. It can only be reached in the temporal present. Like the ethnographer, the phenomenologist must engage in performative participation in the experience. There cannot be detachment, but must be personal investment. There are many similarities between phenomenology and ethnography, but it is also important to point out that they are not the same. Both attend to the "data"[4] of experience and often will use similar methods to collect the data. Both rely on reflection for analysis. However, there is a "radical reflection" in the phenomenological method's use of the *epochē*. Also, there is a difference in the location for phenomenologist. The ethnographer will remain in the world and stay at the level of pure description. The phenomenologist will be interested in how the descriptions come to be held in consciousness, how they become sedimented as part of experience, and the essential structures of intentional consciousness that the data point to. So, there are many similarities between a certain type of ethnographer and phenomenologist, and they could even share the same "research design." However, the methods require different types of reflection and ultimately move in different directions.

To practice the phenomenological method, the first step is to get immersed in the natural attitude and lived experience of the phenomenon to be investigated. The phenomenologist should participate fully in the event. They should experience the phenomenon as deeply as possible. Go to the event and do it. Also, the natural world is a richly social world, so we should attempt to engage others as they share in these experiences. This involves rich observations and conversations. Talk to people, listen to their stories in an open way as a fellow participant. Collect as many stories as possible and be aware of the environment around where the experience is taking place.

Phenomenology of Visitor Narratives

I spent hundreds of hours interviewing hundreds of zoo visitors both formally and informally talking about their particular motivations for visiting the zoo. I asked them questions like why they came to the zoo that particular day. What was the purpose of their visit? What did they hope to do? Why did they come to the zoo instead of somewhere else? This was buttressed with my own family visits to the zoo. The idea is to become immersed in the experience. This is the natural world of the taken-for-granted every day that needs to become clear in order to enter into the phenomenological method. After all we need to know what it is we will be bracketing.

The purpose of these interviews was not to gather data to make an objective claim about zoo visits. Instead, the intention was to try to see what reasons the visitors gave for going to the zoo. I was not interested in the quantitative assessment of how many people gave a particular type of answer. Instead, the natural world analysis tries to grasp the experience itself as lived. The words of the visitors, their feelings, and assessment are what matter. The "real" objective truth (whatever that means anyway) is inconsequential because this experience is about visitors' perceptions of why they came to the zoo. Therefore, the concern was with the depth and meaning of the stories and not if zoos actually delivered (or could deliver) on what drove the visitors to the zoo that particular day.

Capturing these stories is problematic because it takes visitors out of the lived experience of the visit. It can be seen as intrusive, and therefore many visitors rightfully won't stop to be interviewed during their visit. They also might not complete surveys, even if offered a free drink or some other reward from the zoo. This reluctance of the visitor to break from their visit and their time must be respected by the interviewer. The researcher is an intrusion that can change the dynamic of the visit. So, the mere asking of the questions changes the nature and experience of the visit. This must always be kept in mind.

One final word on method: the phenomenologist is ultimately engaged in philosophical anthropology. Philosophical anthropology is concerned with the domain of human experience as it is lived in the context of the world. There is an "ecological"[5] focus necessarily in this analysis. There ultimately will be a concern with the context and history in which these meanings arise as part of this methodology. The phenomenologist is concerned with how meanings both passively and actively come into being in their full, rich context. The starting point for this philosophical anthropology is the experience in the flow of everyday life.

In this next section, I attempt to group together some of the different motivations for visiting the zoo. The purpose of this list is to give a qualitative assessment of visitors' perceptions of the zoo visit *not* to say that zoos actually do this or that. These are the lived experiences of various visitors. In order to make sense of them, I have grouped them together to help the reader make sense of them. Husserl understood that experience has an underlying structure. Therefore, the groupings are meant to help find shared structural experiences of the zoo visit across various groups of visitors. I am attempting to make the grasping of the essences of the motivation of the visit more transparent in the grouping and make the narratives of experience easier to access for the reader. However, in these groupings some degree of the individual uniqueness of each empirical experience is lost. This is done though because there is a belief that there are shared structures to experience shared, just by being human. I cannot actually "feel your pain," but I can access similar experiences of pain in my personal experience that I can "pair" with my perception of your "pain" that can help me empathize with your experience. To be human is to have the same capacity for experience based upon a shared essential structure in experience. Otherwise the capacity for understanding, communication, and empathy would be impossible. We all are human and as humans engage in experience with a similar, sedimented structure of experience.

This may seem like a controversial proposal in the postmodern wake of Foucault who claims the modern human subjectivity will be "erased, like a face drawn in the sand at the edge of the sea" in favor of radical difference.[6] Foucault's "death of man" concern was with the modernist tendency to reduce the human experience under a colonizing domain of power in the context of human consciousness. Foucault objected to the Enlightenment's effacing of difference and instead turned to language. The phenomenologist shares Foucault's concern with the colonizing reduction to a unity of sameness engaged in by modernity. However, phenomenology rejects Foucault's critique of the rational subject and, instead of archeology, suggests a radical quest to uncover origins. The problem with the washing away of the modern subject from the point of view of phenomenology is that it allows for a very detached or disembodied notion of anonymity. Strangely enough for someone who writes on power, this ignores the particularized embodiment of the oppressed. As Lewis Gordon has pointed out in *Her Majesty's Other Children*, the washing away of the subject creates a strange anonymity that allows for discussions of white supremacy without ever referencing white supremacists and allows for a neoconservative stance that there is no such thing as race.[7] And as Martin Matustik has pointed out there was "a general failure to see the faces that stood behind texts and textualization."[8]

The phenomenologist, instead of rejecting the rational subject and science under modernity, attempts to save science and the subject from crisis by making origins transparent. For a more detailed, critical assessment of the postmodern rejection of science from Foucault in favor of a phenomenological approach, see Calvin O. Schrag's fantastic work, *Radical Reflection and the Origin of the Human Sciences* (1980).[9] To get into the details of this critique is beyond the scope of this project and I could not do justice to Schrag's argument in this space.

VISITOR MOTIVATIONS FOR ZOO VISITS

In this section I offer nine distinct categories for making sense of the zoo-visit motivation. I believe that these could be a good starting point for anyone wishing to attempt a more empirical study of zoo-visit motivation. Keeping these categories in mind could help generate some survey choices, structure oral interview questions, and/or categorize visitor narratives.

Entertainment

One of the most common motivations for a zoo visit is entertainment. When asked why they are visiting the zoo, visitors with this motivation will often respond: "for fun," "to enjoy the day," "to have a good time," and so forth. For these visitors there is something pleasurable in the zoo-visit experience. They come to the zoo as they would go to the movies or engage in some other enjoyable experience. It is fun for them. The visit offers a thrill. This entertainment experience is often connected with children. Children will talk about how much fun the zoo is, and parents will mention that the motivation for the family trip was for the enjoyment of the children.

Zoo directors have acknowledged the recreational motivation for zoo visits for quite some time.[10] However, this is always a troublesome acknowledgment. There is a worry that "mere entertainment" is not enough of a motivation for the maintaining of a zoo. Zoos want to aspire to more. These recreational motivations are "prejudged" to be lesser motivations. At this stage in the phenomenological analysis we must avoid prejudging these recreational experiences. Yes, visitors come to the zoo to have fun and if a zoo is thought of as just being a collection of cages for animals that increase their suffering then the mere entertainment motivation would be deficient.

Yet, many zoos are designed in ways that reinforce these entertainment feelings. For example, many zoos share multiple design features and layouts with amusement parks. Before you even enter the zoological park you enter a large parking lot. Then you go to a gate where you purchase tickets to enter the zoo. There are concession stands found throughout both type of parks. They both rely on guest service employees to sell us very similar food and merchandise. The zoo and the amusement park often compete for the same entertainment dollars in family expenditures and thus use publicity that sell them as "family entertainment." Zoos and amusement parks have similar family crowds waiting in lines to go through different experiences. For the recreational visitor there is something similar in the experience of riding a roller coaster or walking through an immersion exhibit in the primate house.

What are some of the characteristics of this entertainment that make it different from other forms of recreation? First of all, the predominately visual nature of the experience of the zoo visit must be acknowledged. The zoo visit relies on visual stimuli of the animal in the exhibit to provide the entertainment. The enjoyment in the zoo is experienced primarily through the visual stimuli. In this way we could say that there is a metaphorical similarity to how the entertainment is achieved with the same pleasurable enjoyment of going to the art museum and looking at a painting or sculpture. I do want to point out that the experience is not entirely visual. There are very unique smells, sounds, and tactile experiences that also characterize the typical zoo visit.

Yes, the zoo visit is fun for many. And for many families this entertainment is the primary motivation for their initiating a trip to the zoo. Therefore, zoos must keep positive associations around the experience of the visit to keep their guests coming back for more. That is why negative experiences like a messy exhibit, too large of a human crowd, being too expensive, not enough to see, and so forth, can take away from the overall pleasure of the experience of the visit, lead to a negative evaluation of a particular zoo visit, and cause visitors to avoid a particular zoo.

Relaxation

Another common response for visiting the zoo is "to relax," "to escape," or "to get some peace." This response has many similar qualities and characteristics as the entertainment motivation. However, there is also something substantially different about the "relaxation" response that calls for a separate analysis.

One way to view this response of "relaxation" is in line with entertainment—that we humans lead busy, modern lives and that we look for moments to escape the hustle and bustle and relax. From this perspective, relaxation is an enjoyment that is sought in response to the work of the everyday. However, another view of the relaxation response is that there is something inherently "peaceful" in the zoo experience for many individuals that causes them to go to the zoo.

It is no accident that many zoos are found in parks. The evolution of zoos from "menageries" to zoological parks at the turn of the twentieth century will be examined in chapter seven of this book. In the United States and Western Europe most zoos are located in major cities. They were formed in large, public parks and spaces. These would be the places urban citizens would go "to be in nature." Many zoos maintain large arcades of grass, wooded areas, or even streams and marshes on their grounds. The grounds of many zoos are landscaped to present neat and controlled spaces, yet at the same time there is plentiful use of various flora to complement the immersion into nature and out of the hustle and bustle of the city.

Why do we turn to living plants and animals for relaxation? A number of empirical studies have been done to show that there is a direct health benefit to humans from viewing and being with other living creatures.[11] Having a goldfish in the dentist office has decreased the need for anesthesia. Enjoying the companionship of one's pet dog has been shown to improve one's recovery from various heart ailments. So, there seems to be something that we gain from viewing and being around animals. It causes a physiological change in us that can best be described as making us "at peace," or helping us relax. Psychologists have even recognized this soothing effect and for this reason some have used "companion animals" to help patients open up during therapy. The dog as comforting elicitor of "intimate dialogue," capable of helping children open up and express themselves, has been observed since the 1960s when psychoanalyst Boris Levinson noticed how having his dog in the room during a therapy session helped a very reserved and introverted child open up and talk. Levinson was the pioneer of animal-assisted psychotherapy and now some therapists seek the help of dogs to horses, and a variety of creatures, because they seem to naturally provide an opportunity for children to open up and speak due to the relaxed state invoked by the animals.

The presence of the animal creates a protected more soothing environment that is conducive to the therapeutic effect.[12] Companion animals can even have a beneficial effect on linguistic therapies for speech disorders. The *Diagnosis and Statistical Manual of Mental Disorders (DSM IV)* even "notes stuttering is often absent during oral reading, singing, or talking to inanimate objects or pets."[13] So, the child needs a comforting environment in which they can practice speaking and not worry about being judged or ridiculed to help get over their impediment. The naturally comforting environment that is part of the horizon of the conversation with pets can be a useful tool for providing children opportunities to practice their speaking with reassurance.

In a *60 Minutes* interview, Tiger Woods even noted his dog was part of his speech therapy as a child to overcome a severe stuttering problem.[14]

There has been no empirical research done on zoo viewing and human physiology, but if viewing of live animals in close proximity causes calmness in various people, we could hypothesize about the possibility. As a matter of fact, the going to the zoo to "relax" response tells us that visitors are visiting zoos because they anticipate such a positive soothing effect. It would be interesting to see if certain animals have a stronger relaxation effect than others. Do polar bears displaying repetitive behaviors increase or decrease our sense of relaxation? Does the our closer proximity to the animal on display increase the relaxation? These would be very interesting physiological investigations to conduct. However, the empirical data is irrelevant to this phenomenological analysis. The point at this particular moment is that many people come to visit zoos all over the world as part of their relaxation.

How is it possible that just sitting at the aquarium and watching the fish swim can lower our blood pressure? How does this perceived "relaxation" effect operate? Since this is still a natural world reflection let me offer one hypothesis. I think this says more about the sad state of our modern lives full of stress and overstimulation than anything else. The relaxation that comes from viewing or touching an animal pulls us momentarily out of the normal flow of our noisy lives and harkens back to simpler evolutionary times.

Another hypothesis I have has to do with the human relation to time. Our lives are increasingly measured by more and more temporal devices. Our computers have clocks on the corners, we wear watches, we carry cell phones that remind us of time: the measure of time is constant in our lives. There is something anxiety-producing in this temporal fixation. It is as if we are the anthropological rabbit in Lewis Carol's *Alice in Wonderland*, obsessed with "being late for a very important date." Yet, this temporal obsession is not a problem of rabbits or other of our animal kin. We constructed these temporal measurements and are responsible for our own enslavement to the second hand. Yet the animals in the zoo and aquarium don't have this same relationship with the temporal and measurement that we do. Perhaps the relaxation that comes from communing with animals in the zoo is a direct result of this reminder of an idyllic temporal moment.

The relaxation motivation for a zoo visit is similar to the motivation for doing yoga. The zoo causes a physiological response in the visitor that makes them more calm and peaceful. There is a tranquility that is being sought and for many guests this is found in the experience of the animal. Watching animals is a technique of relaxation that decreases the visitors perception of stress.

Aesthetics

> Tyger! Tyger! burning bright
> In the forests of the night,
> What immortal hand or eye
> Could frame thy fearful symmetry?
> —William Blake[15]

Zoos provided many people animals to see that they otherwise would never see during the normal course of their lives. Painters, photographers, poets, writers, sculptors, and other artists have visited zoos to find inspiration and models in the various animals on display. Early on, this inspiration for artists even factored in as part of the argument for establishing a zoo. "The usefulness of zoo animals as models for the drawings of artists and naturalists was a common justification for establishing a new zoo. Animal art also served the educational goals of zoos."[16] Those arguing for the construction of a zoo would explicitly state the zoo would provide a source of inspiration for the local community of artists.

One of the most common activities that visitors engage in is photography. Even exhibits explicitly marked as not allowing photography have to deal with numerous amateur photographers flashing away despite posted signs. I have approached several visitors and asked them why they took a particular photograph and the answer is usually something along the lines of "the tiger is just so beautiful," or "I just find the colors of that snake so amazing." The photographers hit a key zoo-visit motivation identified by zoo visitors. There seems to be something aesthetically pleasing about viewing animals. "Animals are pretty," or "we came to see the beautiful animals" is a common refrain among many visitors. The wows and awes of a crowd of visitors indicate a perceived "majesty" in many of the animals on display at the zoo. The colors, textures, markings, symmetry, and other visual stimuli of various animals offer guests an experience that they equate with being pleasurable. For many guests this beauty is indicative of the aesthetic association with many animals.

I do want to point out that there is even a reverse aesthetic association that should be investigated. Even animals such as snakes, Madagascar hissing cockroaches, bats, spiders, or naked mole rats have regular visitors who are attracted to regularly viewing them. Oftentimes you can hear children shout, "Ewww gross," or a mother shriek, "That is disgusting." Yet, zoos continue to display these animals because they offer a unique motivation for visiting. It is good to have our aesthetic sensibilities challenged. A few zoos have played with signs and advertising featuring these animals and captions like "a face only a mother could love." But some guests do find a beauty in these animals. While others come to view the animals just to be repulsed.

What does it mean to find a particular animal beautiful or ugly? The animals are symbolic of the sublime beauty and stark ugliness we find in the world. Aesthetics is tied to human nature. We search for beauty, and even sometimes the ugly, because having aesthetic experiences are considered an important part of our daily lives. We develop and refine the particularities of our aesthetic judgments in a cultural context, so different animals often have a different aesthetic resonance in different cultures. Often subconsciously, we will structure our lives and make choices based merely on preferences for different aesthetic experiences. The zoo animals offer a vast slate of various stimuli.

One of the more interesting studies I was a part of involved discussions with children about what they thought "animals needed to be happy."[17] Since we recognized that children did not have the rich language to often fully describe the complexity of their views of animals, we decided to engage the children in a dialogue through drawing. A researcher would ask children what their favorite animal was. The child would be asked to draw the animal. Then the researcher would ask "if the animal was happy" and, "What does the animal need to be happy?" The child would then draw what they thought the animal needed to be happy. We conducted these drawing interviews with 171 children ages four to fourteen. We were surprised by one particular response that pointed to similar category of response. A girl drew a zebra as a favorite animal of hers. Then she started drawing flowers for the zebra to be happy. She was asked why the zebra needed the flowers. The girl was insistent that they were not for the zebra to eat, but that the zebra needed the flowers because "they were pretty." In the follow-up to why zebras and animals needed pretty things, the child said because she likes to be around pretty things and so does the zebra. Other children drew grass, trees, and other elements not for food, but to complete the natural surroundings of the animal. These children responded that these elements were not for the animals to eat, but were there because the animal needed them to be happy. The happiness for the animal was caused because these items completed the beautiful environment that the animal enjoyed.

Other children talked about the "bodily coherence" of the animals. That part of an animal's happiness was dependent on being a "beautiful, whole, lion," or a "complete penguin." One could have just dismissed this response as being a form of anthropomorphism on the part of the child. However, we had several responses in which the children emphasized that animals needed "pretty things" or "beautiful things." There is something different in these aesthetic characterizations from the lower, more direct anthropomorphic responses like the animals' needing hot dogs, Mountain Dew, or Power Ranger toys. Some children identified animals as having a desire for aesthetic elements just like we do.

Education

If you want to have a wild time observing visitors at a zoo watch and listen to their behavior in front of baboons while they are in heat. During this time females rumps will swell and become more engorged and red than usual. One of the funnier (or sadder) interactions I observed was a child about five years old asking his father what was wrong with the monkey's butt. The father responded, "he has hemorrhoids." You would be shocked how uninformed some *adult* visitors are about animals. I have heard animals repeatedly misidentified, as the visitor stands right in front of the sign explaining precisely what the animal is. Or adult visitors completely misunderstand the workings of basic physiological processes. Someone came over and told us that "the little pigs were eating the big one." We went over to the warthog exhibit where the person pointed us and found three piglets nursing. If there was a book of the stupid stuff that people say at zoos I would be the first to buy it.[18]

Many visitors state that their reason for visiting the zoo today was to "learn something about animals." There are those who are curious about the animal world and come to the zoo to learn about various animals. For example, I heard a mother interviewing a keeper in the free flight bat exhibit in the Australia house about what she could do to make her backyard more attractive for bats to help her keep her mosquito population down. Or I have seen children become encyclopedias of information about zoo animals as they play zoo veterinarian. Wanting to "learn about the animals and their habitats" is one of the more common responses for zoo visitors explaining why they visited the zoo. The educational motivation is one that zoos have recognized for a long time and have used to justify their creation, and is ingrained in many zoo and aquarium mission statements.[19] Many zoos have education departments, teach classes on their campus, provide docents to help deliver information to visitors, and keep and maintain websites with various facts about the animals kept at the zoo.

Despite the educational motivation and intentions of visitors, the success of the educational experience varies. For example, many visitors will not read exhibit signs, or will merely skim them over. There is not a lot of research on the educational impact of zoos and aquariums on visitors. My findings mirror David Churchman's 1985 essay in which he attempted to codify a method to investigate what information visitors are exactly taking away from zoo exhibits. "When interviewed, visitors are neither surprised nor do they object to questions concerning what they have learned. Rather, all readily accepted the assumption that they should have learned something."[20] When I have quizzed visitors on the information from the signs they have just appeared to have read, I have been surprised by the variety of responses. Some visitors have an almost photographic memory, others miss a few minor points, others remember little-to-no information. The amount of retention of educational information is also limited by the amount of knowledge that the visitor is bringing into the zoo. It is true that many visitors already possess the basic information presented on many signs presented at the zoo.

Zoos also have to be careful that they don't provide misinformation. For example, visitors may take the "abnormal behavior" of a particular animal as indicative of an entire species. This animal may be lethargic in this exhibit on this particular day, but that does not mean the entire species behaves this way. Or abnormal physiology could confuse children or visitors. A particular animal may have had an operation to save its life and could have had a limb removed due to an injury. A visitor could become misinformed about the animal based on exposure to this animal. Also, enclosures become a "teachable moment" about animal habitat.

The educating of the public is of a central importance to zoos. Education is a common justification given for visiting the zoo or bringing one's children to the zoo on any particular day. Yet, we have reason to be skeptical about what people are actually learning from the animals. When a family happens upon a pair of camels mating and the mother explains that the "animals are fighting," are we protecting our children from some "age inappropriate fact" or are we "misinforming our children"? Some people seem to have a desire to "learn about animals." This may explain the plethora of animal and nature documentaries. As a matter of fact, over 30 million people turned in to view the 2010 Shark Week on the Discovery Channel.[21] Sure people tune into these shows for entertainment, but also to learn. The balance of the amount and age level of the information to present to visitors is a challenge that zoos and museums are constantly struggling with. Since visitors are coming into the zoo with educational levels from preschool to professionals with advanced biology or zoological degrees, what is the appropriate amount of information to present? How can an exhibit sign hold the interest of the PhD in zoology and the elementary schoolchild who has just learned to read? And how can this information be presented to a mass audience as succinctly and clearly as possible so that it avoids confusion? Also, how can we maximize the information that visitors can take from their visit? This is a challenge that zoos are constantly struggling with.

Social

Another common reason given for visiting the zoos is social. This can be divided into two distinct types of socialization motivations. There are those who go to the zoo to be with their friends and family in a social experience. Then there are those who describe a motivation for going to the zoo as a social one to be with the zoo animals. The zoo visitor can give a motivation for bonding with humans or animals.

We mentioned earlier that the typical zoo visitor group involves family groups of parents with children. Many explicitly mention wanting to spend time with their relatives and friends as being a central motivation for their zoo visit. "I went to the zoo to spend time with my kids," or "we wanted to show Grandma the baby tiger." Going to the zoo is a social event for families and various groups of people. The zoo visit is a way for multiple generations to spend time together and engage in a shared event. There is a shared goal of visiting the zoo. People are brought together in close proximity to do the same activity—view the animals in the zoo. School groups also share a bond when they bring grades of children to explore the zoo together. Going outside of the classroom and travelling together reinforces social bonds and creates new friendships. I have even interviewed young couples who came to the zoo on a "first date." I remember asking a particular couple why they came to the zoo. They mentioned that they both expressed an interest in animals and thought this would be a "different way" to learn about each other.

The zoo can strengthen our bonds with our human circle of friends. The zoo can also provide an opportunity to socialize with animals. A number of guests have mentioned that their primary motivation for visiting the zoo that particular day was to "be with the animals." When they were pushed on what that means, they would mention that they want to feel a connection with the animals in the zoo, so they come to visit them. The metaphor of "friend" or family member that humans use to describe companion animals in their house can also be used to describe zoo animals. As a child I felt that Olga the walrus was "my friend." She had a recognizable personality. I wanted to spend time with her because to me there was a connection that she provided to the world. There was a hard-to-explain intimacy in my visiting her. I knew her name, studied her behavior, I would spend the most time with her, and would even talk to her. This social connection with animals is not a strange occurrence described by children, but also something engaged in by adults. Adults would say that they came to the zoo to "be with the wolves," or to visit a particular polar bear, too.

To be human is to be a social animal. We make connections and the zoo offers a place for creating and maintaining bonds with the nonhuman world. These bonds are not necessarily only for lonely people without human friends. Instead, the animals offer us a supplement and different type of bonding. And the place of the zoo offers a unique situation in which these bonds can be formed.

This social function relies on emotions to establish the connection between human and the animals at the zoo. In their pioneering research on zoo animal viewing and spontaneous affective emotion, Myers, Saunders, and Birjulin give three specific functions for emotions. These are:

- "The *indicative function:* Emotion as a cognitive system"

- "The *mobilization function*: Arousal and emotion regulation"
- "The *communicative function* and socio-cultural context"[22]

Indicative Function

The indicative function states that the emotions function beyond our reflective, and intentional (as well as phenomenological) state and provide an evolutionary advantage by allowing us to unconsciously react to the world in a positive way. The indicative function allows an "automatic" affective emotional response to the stimuli in the world. Emotions help us by giving a quick reaction to danger and not having us necessarily engage in rational appraisals that require a longer, cognitive deliberation.

Mobilization Function

The mobilization function of emotion points to the connection between the mental and physical. Certain emotional states trigger physiological responses. Our bodies have a physical response and literally move in response to the effect of various emotional states we achieve in reaction to various phenomena in the world.

Communicative Function

The communicative function explains emotion in terms of the social interactions we have in the world. Emotions provide an experience to help us regulate our social interactions with other humans and animals in the world. Affective emotion functions to motivate us to connect, or disconnect with various animals. It stimulates communication and interaction and thus is a major impetus for social bonds.

Going to the zoo then provides people the opportunity to feel a connection with various animals and establish a perceived social bond. One of the more fascinating conclusions of the Myers, Saunders, and Birjulin study was that the emotional bond and social connection was directly influenced by the particular animal and the level of activity the animal was engaged in. Certain animals tend to induce particular emotive responses: for example "fear," from snakes. And how active the animal was in its enclosure also impacted the emotional connection that the visitor perceived to have with the animal. The more active the animal was, the stronger the emotion that was stimulated and, potentially, the stronger the social bond that was formed.

One final note about emotion: many adults recall very powerful emotional experiences with particular animals in the zoo during their youth. They often equate these experiences as being "life changing," or "shaping the way they related to animals" later on in their life. Early powerful emotional experiences can shape later views on animals and an individual's general bonding with nonhuman others. The visitor self-report behavior change based on these "powerful" experiences.

Caring

A particular emotive response that deserves its own attention is "caring." A number of visitors mention coming to the zoo because they "care about the animals." This is a particular emotive response that deserves special attention. In a future section we will discuss the evolution of zoos. However, at this point I would like to note that zoos see themselves as having evolved from simple menageries arising out of modernity to todays "centers of conservation and caring."[23] The holy grail of zoo visitor research is to provide a direct correlation between "caring for animals" and zoo visits. As former director of Brookfield Zoo, George Rabb states,

> [W]ords conveying different forms of engagement with the natural world are also gauges of biophilia: caring that . . . (intellectual), caring about . . . (emotional), caring for . . . (active behavioral expression). Although these terms seem to encompass the natural dimensions of caring, which we regard as the essence of conservation (Rabb 2001a), there are other considerations.[24]

I will discuss the important notion of biophilia later in this book. However, it is important to note that zoos see caring as a powerful emotion that they hope to nurture in their visitors. The belief is that by developing any feeling of caring for an animal, a basis or motivation for conservation is created. Conservation as an explicit motivation in and of itself will be discussed in the next section of this chapter.

I want to pull out this notion of caring and develop it further. When I have asked people what they mean when they say that they care about animals, a common response is that "I care about animals because they are like us." What the guest means is not that the animal is viewed as anthropomorphic, but instead that they have empathy for the animal. They literally care for the animal because they feel that they share many of the experiences that the animal does. This is a curious response from a visitor freely roaming around a zoo where the animals are kept in a state of "captivity." This statement of caring for an animal that is "like me," shows that the guest recognizes the mental and emotional life of animals. Caring also reveals a very intense emotional relationship. We often use the notion of caring in conjunction with human "loved ones." To care for the animals means to feel an intense closeness with the animal.

Caring relationships should be balanced with the idea of zoophobia that is common among many adults and children. Zoophobia is having an intense fear, repulsion, or even hatred for particular animals. Bats, spiders, and snakes are animals that trigger the most fear in people. I recently saw a three-foot snake in my backyard and went to my state department of natural resources to help identify it. I identified the snake as an eastern ratsnake based on the photographs and information provided. However, I was shocked to read on the website that these snakes are often "senselessly slaughtered out of ignorance and fear. The fact is that eastern ratsnakes are one of the state's most beneficial and splendid reptile assets; they play an essential role in controlling destructive rodents."[25] Through fear, stupidity, ignorance, or other motivations we often don't think about or care for particular animals. If we could teach people to "care for" the eastern ratsnake we could naturally control our rodent populations in large cities. Yet, seeing a snake that could reach up to 72 inches invokes such fear that people kill it and remove it when it has a positive benefit to our natural environment.

Zoos hope to impart an "ethic of care" in their visitors. Some visitors come to the zoo because they literally care for particular animals in the zoo. This then translates to other caring behaviors and even caring for other humans. In abnormal psychology research, there is only one common factor in the past histories of serial killers: At some point they probably engaged in animal abuse. We should head Margaret Mead's warning, "One of the most dangerous things that can happen to a child is to kill or torture an animal and get away with it."[26]

As our society becomes more mechanized we are losing the means of connecting to and thus caring for the living world around us. However, it is these connections that are essential to the meaning of what it means to be human. Caring for zoo animals, plants, and pets can be particular points to develop roots of caring for the world around us. This leads us to a larger conservation ethic and concern, not for only a particular animal, but the whole world around us.

Conservation

When the emotional connection of care is transferred from the particular animal to the environment and all animals, this is a conservation motive. Many visitors state they come to the zoo because they support conservation, or want to teach their children to care for the environment, or they want to support the mission of the zoo to protect animals and the environment. Zoos often have conservation and ecological research as part of their explicit mission statements. Some visitors report coming to the zoo and supporting the institution because they identify with the goals and mission of the zoo as a "conservation center." These people might respond with information from a press release about the type of conservation research that the zoo and its staff support around the world. Coming to the zoo is a way of lending support to the institution and to learn about the work done for animals, species, habitats, and environments around the world. Many zoos partner with researchers in the wild, or send their own staff to care for animals in the wild. There is a reciprocity between the zoo and particular species. Zoos need to display unique animals that capture visitor attention. Yet, if these animals become extinct in the wild the ability of zoos to maintain the genetic viability of the species becomes next to impossible. In various visitor surveys, visitors most commonly point to protection of endangered species as one of the zoo's roles that they most actively support and want to know more about.

At this stage of the visit, the move is away from the motive caring for particular animals to caring about entire species and ecosystems full of different animals. Here someone might state they want to come to the zoo because they support the zoo's mission to protect polar bears losing their habitat due to global warming, or orangutans endangered by urban encroachment into their forest habitats. The zoo (rightly or wrongly) is seen as a place to learn about the problems impacting our larger biosphere and to do something to ward off potential extinction of numerous species threatened by human behaviors around the world. These people recognize the precarious position many animals around the world find themselves in. The zoo is a conservation center that offers a potential hope through research, education, and advocacy of caring for the world. Now it is important to know that some zoos have been better at transformation to "conservation centers than others," and that this might not be a visitor motive response in zoos operating under the modernist "menagerie" paradigm, or mere "zoological park."

To clarify, let me give an example here of an interview with a family that travelled through The Swamp exhibit in Brookfield Zoo. The Swamp is an immersion exhibit that simulates local wetlands and teaches about the valuable role that wetlands play in our ecosystem. Wetlands clean the water, slow erosion, and even prevent flooding. They play an important natural role in nature beyond providing home and food for a plethora of animals. Unfortunately, for centuries the value of wetlands were not recognized and in many places they were drained to make way for residential homes, development, or even agriculture. The Swamp exhibit in the zoo presents this information with local animals (including some rehabilitated injured animals) while teaching the important lessons of the wetlands. The family that I interviewed after going through the exhibit said, "This is why we come to the zoo. We want the zoo to teach us how to help the environment.

People will come to zoos to learn best practices for environmental behaviors. Many people will choose what fish they eat tonight based on the guide (or app) from Monterey Bay's popular "Seafood Watch" handout. This sustainable seafood guide actually causes behavior change. The zoo or aquarium provides advice for the visitor to be a successful "steward of the environment." The zoo visit is more than just learning about the environment, or caring for a particular animal. Instead, the zoo visit becomes a guide for behavior change and a way to support conservation activities in one's daily life choices. The zoo also offers visitors coming to the zoo and supporting the individual zoo a connection to bigger and larger support of conservation projects around the world. On zoo websites, newsletters, and onsite signage, information on various conservation activities is given often with the opportunity to give money for support.

Other

The final area of possible motivation I want to mention are outliers. Personally I have not had experience with either of these motivations, nor have I interviewed visitors who admitted to such beliefs. The two behaviors I want to mention are sadism and suicide. In his essay, "Through a frame darkly: A phenomenological critique of zoos," Bernard E. Rollin talks about being a "seriously depressed and physically debilitated" graduate student who thought he was visiting the Bronx Zoo to cheer himself up by seeing the animals. However, his later assessment was,

> I was drawn to the zoo in my darkest moments, I realized that what was going on psychologically was nothing like being made happy. It was, rather, more like gloating over the plight of some beings who were far worse off than I was. . . . [T]he animals were eternally trapped, without even hope of escape. What I thought of as joy was thus better characterized as spite. I understood my situation; however bad it was, I could hope and even anticipate better times. The animals, however, were perpetually doomed to their suffering, with no escape, no hope, no anticipation.[27]

According to Rollin, the zoo is a depressing place and the only true experience to be gained is a form of sadism in which the visitor takes pleasure in directly viewing the pain of the animals. Now granted, I have had conversations with a few visitors who have been concerned about the size of an enclosure for a particular animal, or whether a particular animal was "happy" at the zoo. However, I do not feel that the average visitor views the animals in the zoo as suffering and, furthermore, comes just to see this misery. This does not explain the large and repeat attendance at zoos around the world. Surely reveling in the pain of others is not something that most people seek out and continue to want to see repeatedly. If so, then the worst war zones of the world would also be the top tourist destinations for vacation travelers. However, I do want to concede that there are potentially a few sadists out there who may come to the zoo with the sole motivation to gain pleasure of watching animals that they perceive to be suffering.

Another strange phenomenon worth mentioning is a number of people around the world annually attempt suicide via a zoo animal. These people are obviously not your typical visitor. However, there central motivation for visiting the zoo is to commit suicide via a zoo animal. A visitor might try to swim with sharks, or jump in with the polar bear knowing full well that this will lead to death. Sometimes this suicide may be "unintentional," like when a mentally disturbed individual climbed into our rhino exhibit because he said the "rhino told him to do it." Or the fraternity boy who climbed two barriers and attempted to pet a leopard. Others go to the zoo intentionally to commit suicide via an animal. Around the world people have thrown themselves to animals such as bears, lions, and sharks in order to intentionally end their own lives. These sadistic and suicidal motivations are outliers and do not represent the average zoo-motivation experience.

The purpose of this chapter was to lay out potential motivations that visitors mentioned for wanting to visit the zoo. This list is offered as an initial foray into potential reasons guests may come to the zoo. This is an attempt to describe the immersion experience and get to the "thing-itself," the phenomenon to be analyzed without any rich phenomenological reflection. However, this is valuable because it offers us a glimpse into the most common prejudgments visitors hold with them before they go to the zoo. These visitor perceptions are important to deal with because they impact the experience that the zoo goer will have. For example, if the family comes to the zoo for entertainment and the zoo bombards the guests with information about conservation practices and behavior change, the zoo may lose this guest, or this group may have a negative experience and not really know why. Also, zoos need to have clear missions and be clear on their own purpose that they communicate with the public to avoid confusion or sending mixed messages. If zoos are to be "conservation centers," then having rides, and being overly inundated with selling merchandise following the design of amusement parks can obfuscate their mission.

These categories also are not mutually exclusive. A family can come to the zoo to socialize with their family and to engage in caring by forming empathic bonds with particular animals in the zoo. So just because an individual comes to the zoo "to have fun" doesn't mean we can't expand their understanding of conservation and move them toward behavior change in their lived daily practices. For the phenomenological method, the point of the natural world assessment is to position the starting point from which analysis will begin. What are the beliefs that are prejudicing the experience? Beginning in the natural world is a participatory way to open up the phenomenon.

NOTES

1. Many ethnographers were phenomenologists (such as Schutz, and Garfinkle), there were ethnographers who studied phenomenology (such as Geertz, Bateson, and Mead), and finally there were those who were indirectly influenced by phenomenology developing parallel ideas (such as Levi-Strauss, Malinowski, and Mauss).

2. Clifford Geertz, "Thick description: Toward and interpretive theory of culture," in Clifford Geertz, *The interpretation of cultures: Selected essays* (New York: Basic Books, 1973), 3–30.

3. For information on the KRUNK program, go to www.centeroflife.net/krunk-movement.

4. I do not use this word in the empirical or scientific sense, but in the qualitative, experiential notion of the term instead.

5. By "ecological," I mean to consider the living human being in the context of their social and environmental surrounding communities. The human being is to be considered only "in relation" to other human beings and the world around him or her.

6. Michel Foucalt, *Order of things: An archeology of the human sciences* (New York: Vintage Press, 1994), 387.

7. Gordon Lewis, *Her majesty's other children: Sketches of racism from a neocolonial age* (Lanham, MD: Rowman & Littlefield Publishers, Inc., 1997).

8. Ibid., 104

9. Calvin Schrag, *Radical reflection and the origin of the human sciences* (West Lafayette, IN: Purdue University Press, 1980).

10. Jeffery Hyson, "Education, entertainment, and institutional identity at the zoo," *Curator: The Museum Journal* 47, no. 3 (2004): 247–51, doi: 10.1111/j.2151-6952.2004.tb00122.x.

11. See Katcher and Lynch on dogs, and even Beck on fish. For example, see: Aaron Katcher, Alan Beck, Erika Friedmann, and James Lynch, "Looking, talking and blood pressure: The physiological consequence of interacting with the living environment," in *New perspectives on our lives with companion animals*, ed. Aaron Katcher, and Alan Beck (Philadelphia: University of Pennsylvania Press, 1996), 351–59. There is even a documentary called *Intimate Companions* on this research. See also: Alan Beck, and Aaron Katcher, *Between pets and people* (West Lafayette, IN: Purdue University Press, 1996), 78–84.

12. See Boris Levinson, "Pets, child development and mental illness," *Journal of the American Veterinary Medical Association* 157, no. 11 (1970): 1759–66. And Boris Levinson, "Interpersonal relationships between pets and human beings," in *Abnormal behavior in animals*, ed. Michael Fox (Baltimore, MD: Williams and Wilkins, 1975), 504–22.

13. Alan Beck, and Aaron Katcher, "Future directions in human-animal bond research," *American Behavioral Scientist* 47, no. 1 (2003): 80, doi: 10.1177/0002764203255214.

14. Ed Bradley, Tiger Woods interview on television, March 26, 2006.

15. William Blake, "The Tyger," in *Songs of Experience*, (Mineola, NY: Dover Publications, 1984).

16. Elizabeth Hanson, *Animal attractions: Nature on display in American zoos* (Princeton, NJ: Princeton University, 2002), 135.

17. Olin E. Myers, Carol D. Saunders, and Erik Garrett, "What do children think animals need? Aesthetic and psycho-social conceptions," *Environmental Education Research* 9, no. 2 (2003): 305–25, doi: 10.1080/13504620303461

18. I admit to being a member of an online group of zookeepers who explicitly share and commiserate in the crazy things that they have heard people say while visiting the zoo. This has convinced me my stories are not a statement about the level of education of zoo visitors at a particular zoo, but seems to be a quite universal occurrence.

19. Hyson, "Education, entertainment," 247–51.

20. David Churchman, "How and what recreational visitors learn at the zoo," paper presented at Annual Western Meeting of the American Association of Zoological Park and Aquarium Administrators, Anchorage, Alaska, March, 1985.

21. Robert Seidman, "Discovery Channel's 23rd annual 'Shark Week' is most-watched ever; Seen by 30.8 million people," *TV by the Numbers*, last modified 2010. http://tvbythenumbers.zap2it.com/2010/08/10/discovery-channels-23rd-annual-shark-week-is-most-watched-ever-seen-by-30-8-million-people/59634/.

22. Olin Myers, Carol Saunders, and Andrej Birjulin, "Emotional dimensions of watching zoo animals: An experience sampling study building on insights from psychology," *Curator: The Museum Journal* 47, no. 3 (2004): 299–321, doi: 10.1111/j.2151-6952.2004.tb00127.x.

23. George Rabb, "The evolution of zoos from menageries to centers of conservation and caring," *Curator: The Museum Journal* 47, no. 3 (2004): 237–46, doi: 10.1111/j.2151-6952.2004.tb00121.x.

24. Ibid., 241.

25. Ohio Department of Natural Resources, "Eastern ratsnake or black ratsnake." www.dnr.state.oh.us/Home/species_a_to_z/SpeciesGuideIndex/blackeasternratsnake/tabid/6556/Default.aspx.

26. Margaret Mead, "Cultural factors in the cause and prevention of pathological homicide," *Bulletin in the Menninger Clinic* 28 (1964): 11–22.

27. Bernard Rollin, "Through a frame darkly: A phenomenological critique of zoos," in *Metamorphoses of the zoo: Animal encounter after Noah*, ed. Ralph Acampora (Boulder, CO: Lexington Books, 2010), 57–66, 59.

Chapter Four

Bracketing

A Trip to the Zoo

EPOCHĒ

Randy Malamud begins *Reading Zoos* with the sentence, “I do not like zoos; this book begins from that premise.”[1] I may not agree with his starting point, but this is a very phenomenological statement. He begins with a statement of his bias recognizing that his views do not come from nowhere. Malamud, Dale Jamieson[2] and others of this ilk begin from the starting point that zoos are inherently evil. They feel that animals have an inalienable right to be free and zoos impede the individual freedoms of particular animals kept on display. The potential benefits of education, conservation, and so forth, are either argued to be nonexistent or not worth the violation of animal rights. On the other hand, zoo directors and employees so embedded in the project of contemporary zoos may only see the positive benefits of zoos and would therefore never question the legitimacy of the institution itself.

For the phenomenologist, accounting for one’s beliefs and attitudes is an important first step of doing phenomenology. What are the prisms through which I view the phenomenon to be investigated? In the natural attitude we do not take stock of our beliefs, we just act in the flow of life. So in the previous chapter visitors went to the zoo with one of the seven primary motivations (entertainment, relaxation, aesthetics, education, social, caring, or conservation) without reflecting on where this particular attitude came from our how realistic it is for the zoo to satisfy the particular desire.

Phenomenology is a method and one that is involved in the infinite pursuit of origins. The phenomenological method proper begins with th
The *epochē* is the phenomenological reduction. It is a putting aside

beliefs, attitudes, preconceptions, prejudgments, and paradigms and choosing to look at the phenomenon of analysis in a new reflective light. As pointed out in chapter two, this is not a doubting, but instead is a "bracketing" of our beliefs.

The phenomenological reduction requires a free choice on the part of the individual to hold off judgment. The *epochē* is also an ongoing and "infinite task." Commitment to the phenomenological project is a continual method. The world we live in is a cultured world and one that is undergoing constant revision as new attitudes and perceptions shape and form new beliefs and behaviors. Therefore, the phenomenologist is constantly working to refine and improve the phenomenological method as it is being employed. It is quite possible that this analysis in 2012 will be completely different as zoos, animals, and our cultural attitudes toward them change in twenty years. To be a good phenomenologist one must be open to change.

So with that caveat let us begin the *epochē*. The *epochē* is entered into by an empirical ego. So, the first step of the phenomenological method is to begin with an accounting of one's self. From where do you stand? Who is the person who is attempting the phenomenological method? So, begin by telling your story. Understand how you became the person you are and how these beliefs came to be held by you. This is why I claimed that Malamud began with a partial phenomenological method because he started by accounting for his bias against zoos. For me, I begin on the opposite side. I grew up in a poor immigrant family in the near west suburbs of Chicago. Brookfield Zoo was within walking distance and my mother would often take my sister and me to the zoo to escape the poverty, divorce, and other problems of those early childhood days. I took classes at the zoo. To this day I still remember the way to identify particular ibex based on the shape of their horns. As a shy child it was a place where my "friends" lived. It was a place where I developed and first understood empathetic thinking, often wondering about the internal life of the animals around me. My general environmental sensibilities were shaped in the context of the zoo; wanting to protect the environment and preserve the ecosystems of the other polar bears, walruses, baboons, dolphins, and penguins who were not present at the zoo. After these visits my family became actively involved in dolphin research projects around the world; we engaged in boycotts of corporations whose practices caused unnecessary harm to animals. Even when we had very little, we gave money to groups like the World Wildlife Fund, and actively followed the work of conservationists like Jane Goodall. These commitments and actions were a direct result of having formed bonds with all the particular animals that I visited regularly at Brookfield Zoo as a child. This is my starting point. However, that doesn't mean that I cannot be critical of zoos.

Ziggy

One can be critical of whether an animal has enough space, or concerned about its health and overall well-being. The year before I was born, Ziggy the elephant died. Ziggy was one of the most popular animals at the zoo from 1936 to 1975. Ziggy had been a circus elephant prior to being bought by the zoo. During an attempt to mate him, Ziggy attacked a keeper. Instead of killing Ziggy, the zoo kept the elephant chained in the pachyderm house for almost thirty years. After a local advocacy campaign on behalf of Ziggy, a more secure outdoor enclosure was built for Ziggy, but only after thirty years. Ziggy was only allowed to travel outdoors for the last five years of his life.

Questions about health, safety, well-being, space for enclosure, human expertise, and many other factors must constantly be examined by zoos. The conditions of particular zoos and animal exhibits around the world have a wide variety of forms and some are not appropriate. Ziggy was chained for thirty years because of a human mistake in how he was handled based on insufficient training and knowledge at that particular time. If the keeper understood that Ziggy's testosterone was elevated during his musth period (an increased hormonal time found in bull elephants), the attack may have been avoided. Ziggy should not have been chained indoors for three decades.

The lesson of Ziggy is that perhaps some animals just should not be displayed at zoos. We connect zoos with "megavertebrates" or large animals. Bears, giraffes, gorillas, lions, elephants, and so forth, are the animals that attract the most visitors, and thus lead to the most profit for zoos. Yet, these same animals may also be some of the hardest animals to care for. Perhaps some zoos don't have the space, knowledge, funding, capacity, or other resources to care for particular animals. This doesn't mean that the phenomenological description will be necessarily critically negative. The phenomenologist must leave open the possibility of the assessment either positive or negative and must attend to the thing-itself. So, just as I must set aside my Brookfield childhood experiences, the paradigm of the animal rights held by Malamud, Jamieson, and Singer must also be set aside.

Binti Jua

Brookfield Zoo's Tropical World was built in the 1980s and at the time was the largest indoor exhibit in the world. It tried to simulate the rainforest of three different continents and displayed a variety of animals together in the same exhibit. One enters the exhibit through a waterfall into South America. Then you go into the Asian rain forest with gibbons and orangutans. The final area is the African rain forest complete with a gorilla grotto with western lowland gorillas. Tropic World was one of the first attempts for a contemporary immersion exhibit that removed the bars from the animal enclosure and still is one of the most popular exhibits at the zoo today.

April 16, 1996, was a crazy day at Brookfield Zoo. On that day a three-year-old boy ran to the end of the exhibit while his parents were still at the start. He climbed up on top of a gate above the island where the gorillas were and fell into the exhibit. The boy remained motionless. Binti Jua, an eight-year old female picked up the boy and "cradled him." She kept two adolescent males away from the boy and brought him over to the door where the keepers had access to the exhibit. The boy eventually recovered from his injuries in the local hospital. However, this began a media "circus" at the zoo. Binti Jua was called a hero and was even given an award by then-First Lady Hillary Clinton. Primatologist Frans de Waal even cited the incident as evidence of animal empathy.[3]

Yet, was Binti Jua a hero? What did this incident say about the inner life of apes? What amazed me in the frenzy to point to the similar inner life of Binti Jua to humans was the failure to notice or account for the particular lived experience of Binti Jua in the zoo. Binti Jua had been hand-raised by humans. At the time of the accident she had a one-year-old infant of her own. Keepers had been working with her for several years to prepare her to be a mother. Also, the gorillas were given a reward if they brought objects that fell into the exhibit to the keeper's door. Even among primatologists, little of the ape behavior training was mentioned. Instead, people jumped to conclusions about the "inner life" of the ape claiming to know that the ape was just like us. Claiming this particular empathic understanding of the inner life of animals must also be bracketed. We don't have access to the inner life of animals. Here we should recall the argument against Nagel's "objective phenomenology" in his essay of "what it is like to be a bat" from chapter two of this book. I do not deny the inner life of Binti Jua—that she has the capacity for empathy. Instead, I just believe we need to take up that inner life *qua* ape with a particular history. Binti Jua is a particular ape, with a particular history, raised in a particular context. A phenomenological attending to Binti Jua should begin with her experience, not anthropomorphic presuppositions.

Binti Jua and Ziggy provide lessons about having a pre-established frame in place prior to reflecting on the nature of the zoo. Zoos can be cruel when they do not adequately care for the animals displayed on their grounds. Yet, we must reduce the lives of animals to our own. Now let us bracket the view of the zoo lover and the animal rights activist and visit the zoo as phenomenologist. We will need a new set of descriptions.

DESCRIPTIVE PHENOMENOLOGY

The first step after the bracketing is to try to just describe the phenomenon of analysis in a rich way. Here the ethnographic tool of "thick description" is important. With these descriptions come reflections, and in the reflections we look for structures. What is it like to visit a contemporary zoo in the United States? You walk through the gate with your family. Banners and pictures of large animals such as dolphins, bears, lions, and apes hang in the entrance. Everything prior to this moment is about the anticipation of the animals you will see. The grounds of the zoo are several acres and there will be lots of walking today.

The first exhibit you come upon is a tiger enclosure. From a distance away you see a patch of yellow and black color amongst green grass, brown sticks, and gray rocks. Initially it is this color that catches the eye. There is a bright color that stands out against the dark and earth-colored background. The difference and unusual formation stands out and is "alluring" to the eye, drawing you in. As you approach the exhibit, the color that caught the eye begins to take form. The anticipations of seeing a tiger confirm the shape of a large Siberian tiger lying on a rock in the center of the exhibit. The enclosure is built on a hill and the tiger seems to be eye level with you. Between you and the tiger is a low fence with a step for children to climb on to get a better view of the tiger. There are signs that provide the Latin name of the animal, tell you about its habitat, as well as about behaviors such as eating and mating habits. On the other side of the fence is a deep wide moat that provides the barrier between human and tiger. In the enclosure there are several trees, a few bushes, grass and several large rocks. On the right side is a glass viewing area where tiger and viewer can come face to face. On the left side is the opening to a building where the tigers are kept off view from the public. As you look to the back, you see another tiger lying down near the entrance partially hidden by a bush.

The children in the group exclaim, wow, and shout out the name of the animal. They jump up and down with palpable excitement. Your daughter remarks how the tiger has similar markings to your pet cat that recently passed away. Your son, who is less than a year old, points to the tiger and makes a cooing sound showing interest. As a family, you stand in front of this enclosure and look at the animal. You can detect a faint musky smell from the urine of the tiger. As you stand in front of the exhibit a few other groups have come and gone looking at the tiger. You spend a total of three to five minutes in standing before the animal. The animal does not get up, but does twitch its tail a few times as you watch and wait. You anticipate and hope for the animal to get up, but this does not happen while you are there this particular day. On other days you have watched one of the tigers walking the entire distance of the enclosure from the glass viewing area to the back house. You sense the animal's power in its large claws. You are left with a feeling of awe at the majesty of the animal.

What else can we say about this viewing? It was a primarily visual experience. There was a passivity in the experience in that I was not actively constructing or controlling the event, but instead the tiger was a physical body in the horizon of my gaze. Yet, there was something different in the viewing of the tiger. The rock, tree, or other people did not hold my gaze as the tiger did. So there is something unique about the viewing of the tiger in this context. Perhaps it is the appreciation of the uniqueness of this event. The signs in the enclosure reinforce to me what I already knew: that this tiger's "natural habitat" is half way around the world from here and that the animal is endangered. The viewing of this animal feels like a rare experience. You leave the exhibit with a sense of excitement and explain to your children about the feline family of animals and that this giant cat and the domesticated cat share ancestors with similar characteristics. Then you tell your kids that they are endangered because people have hunted them and we have destroyed the habitat the tigers depend on to live.

Now let us look at the meanings that might be less apparent in that visit. First of all, the exhibit itself is a construction. The layout was not the way patches of this area would be found if there were no zoo there. Every organic vegetation and rock was placed in the exhibit to construct a feeling of nature. Who exactly is this natural feeling for—visitor or tiger? Also, many of the rocks and barriers that appear to be natural are actually made of a natural-looking concrete meant to simulate a rock. The public is not allowed in the back house where the tigers are kept off exhibit. If you did go back there you might find out that there are a few more tigers that are kept in cages separate from the tigers on display in the public enclosure. These tigers may be kept apart to avoid breeding, or because a particular tiger is slowly being introduced to the group for the explicit purpose of breeding under the breeding plan of the Species Survival Plan the zoo participates in to help try to maintain the genetic diversity of the species. However, unless there is an unusual occurrence, these behind-the-scenes operations are beyond the experience and awareness of the average visitor.

The particular tiger I am viewing was most likely born in another zoo somewhere in the United States. Yet, the perceived origin of the animal is a tricky one for the average visitor. The signs talk about a habitat in Far East Asia in the Amur region of Siberia, Russia. Does the visitor know that this animal was most likely bred in a zoo and is part of a select 150–200 tigers bred specifically under controlled conditions to attempt to maximize the genetic diversity of the species? The tiger in the zoo is an amalgam of representations standing in for both wilderness and captivity. We see a captive tiger before us, but the construct of the exhibit and the majesty of the animal itself are meant to signify a wild animal. However, what if this animal had come from the wild, or a circus? What does that mean for the experience of meaning we have in standing before this animal? Even the names of various animals are tied to places. We know of Siberian tigers, African elephants, California condor, Alpine ibex, and the Albanian water frog. The name of the animal itself often directs our attention to faraway places.

The point of this description was not to fall into a relativist framing of the event that leaves it stuck in a divide of individual difference. Husserl's transcendental move was to try to speak across shared experiences to common structures sedimented in the unique manifestation of interpretation. The next two chapters of this book will attempt to pull out two particular eidetic structures unique to the zoo-visiting experience. The first will be to play with the notion of synecdoche as it occurs in the zoo. The next chapter will take up bodily movement in the zoo visit through the phenomenon of children at play with the animals.

PRAXIS PHENOMENOLOGY

What makes this book different is its attention to "doing phenomenology." A lot of the phenomenology done today deals more with the theoretical exegesis of the method, instead of the practical application of the method in our living experience. The question of why do we go to the zoo has forced a shift in the usual focus of the application of phenomenology. The struggle of doing a metatheoretical approach talking about the method or actually doing phenomenology was something that Husserl himself struggled with. The flushing out of phenomenology in various lived communicative contexts is where this project attempts to make a contribution and is what I am calling *praxis phenomenology*. I am intentionally using the term praxis phenomenology instead of "applied phenomenology" because in a real sense all phenomenology is applied since it was meant to engage the lived experience of the world.[4] Also, "applied" often indicates a separation from theory that hopefully my reader has been able to tell is far from my purpose in this project.

Praxis phenomenology begins with a lived experience in context, such as a zoo visit. It uses the phenomenological method to address our theoretical and practical understandings of the world. In my particular use, the communicative and rhetorical meanings take an extra special place of importance. Communication in its manifold uses is a special foundational part of how our social world is structured. To understand these communicative experiences causes a radical change to the world itself. The use of the ancient rhetorical term, praxis, is meant to indicate this life-world altering phenomenon.

As a good phenomenologist I do not want to infer that this praxis phenomenology arises out of nowhere. I consider it to be rooted in the Husserlian tradition. Husserl himself engaged in exercises of praxis phenomenology in his thought experiments watching birds fly or kittens play. I also consider this to have been very much extended in the work of Maurice Merleau-Ponty with his attention to empirical and *Gestalt* psychological research and data while describing his own phenomenological project. Finally, the praxis method of this project builds on the communicative phenomenologists mentioned earlier developed by Schrag, Langsdorf, and Lanigan. Without these thinkers this work would not have been possible.

Praxis phenomenology allows for a particular, focused employment of the phenomenological method with a communicative focus. In the next two chapters you will see particular eidetic structures that are only revealed through a communicative lens. The first chapter will utilize the discourse of rhetoric to look at synecdoche. The following chapter will begin a look at the gestures of embodied nonverbal communication.

NOTES

1. Randy Malamud, *Reading zoos: Representations of animals and captivity* (New York: New York University Press, 1998), 1.
2. Dale Jamieson, "Against zoos," in *In defense of animals*, ed. Peter Singer (Oxford: Blackwell, 1985).
3. Frans De Waal, *Our inner ape* (New York: Riverhead Books, 2005), 3.
4. I credit this point to my first teacher of phenomenology, Marilyn Nissim-Sabat, who would often repeat this point to all of her students.

Chapter Five

Rhetoric and Synecdoche

A PHENOMENOLOGICAL DESCRIPTION OF A CHILD'S ZOO VISIT: PARADOX AND SYNECDOCHE

Let us engage in a phenomenological description of the initial constitution of zoo as spatial object in a child's visit to a zoo exhibit. Prior to the child's visit to the zoo, the animal serves as the *synecdoche* for the zoo in the child's anticipation. A synecdoche is a rhetorical device in which a part stands in for a whole. In our investigation, the zoo is metaphorically represented by animals and usually a particular animal. This "standing in for" is usually of a particular animal. When observing children in various zoos, one does not have to wait very long to hear children talk of "favorite animals." The child's favorite animal usually is the anticipatory representation for the zoo itself in the expectations of a child's zoo visit. These particular favorite animals stand in as representations of the zoo itself in the child's imagination. Ask a child under the age of ten to define a zoo and I believe that many times you will get the answer that it is "a place where animals are kept," and when pressed further they will respond with "the home for my favorite panda—Tai Shan." The meaning of zoo and particular popular animals are intertwined. Zoos realize this and rely on these animals for their branding and promotions.

Habituation

The interesting thing about this synecdoche is that it is predicated on three key preconditions: *habituation*, *individuation*, and *memorable experience*. The first is habituation. The feature of the zoo visit is that many people visit the same zoo multiple times each year. The tendency is also to visit the local zoo more often than ones that are farther away. Therefore, children develop certain patterns of familiarity with their local zoo. Just like my connection with Olga mentioned in an earlier chapter, a child makes very vivid connections with particular animals in a zoo and these are usually the animals of their local zoo. During my time at Brookfield Zoo, when I talked with adults about their most memorable childhood experiences with animals, a good number would respond with incidents and contact with particular animals near the zoo where they grew up.

We are drawn back to the same place and the same zoo. In order for particular animals to stand in for a particular zoo, we must come back to the zoo over and over and have a repetitive experience with a particular animal. Think about a particular animal from your childhood visit to a zoo. Often zoos announce the birth of the animal. There can be a contest and promotion around the naming of the animal. There will be a big event when the animal has its first public viewing. We will receive multiple images of the animal at play in its infancy in zoo literature and local media coverage. Each visit to the zoo will drive us to the animal's exhibit. The particular animal can become part of the habit of the zoo visit—the "must see." Our conscious experience then starts a process of associating the zoo itself with the particular animal that we seek to see each visit. It is a passive process of synthesis that leads to the association of the particular animal with the zoo via habituation. What this means is that we do not purposively revisit the same animal in exhibit in order to associate the zoo with the same animal. This association is a result of our habituation.

Oddly enough, the particular animal begins to habituate into our consciousness as the representation for its entire species as well. Let me give you an example from the pet world. Think about how different each individual type of dog is. Some of us have grown up with great Danes, golden retrievers, poodles, and even Chihuahuas. There is an amazing level of diversity here. Depending upon the particular breed that is most habituated into our memories, when I say dog we will all have very different images of a dog. For me, dog will always be a big, white borzoi. I grew up with Russian wolfhounds and the dog that I shared a house with from my birth to age eleven was Sergei. This gentle giant formed my understanding of what a dog is.

It is important to note that this habituation does not always yield a positive result. For example, my Sergei was always a very friendly dog. His nickname around the house was "Happy" because of his temperament. As a child I would habituate not only Sergei's image, but his behavior as representative of the essence of dog. This could create a false sense of security and safety in my interaction with other dogs. Very often this type of habituation can result in very dangerous child-dog interactions that lead to dog bites. In my case it did as well. Seeing another borzoi when I was five years old, I approached the animal as I would my Sergei, but her temperament and personality were very different. Also, she did not have any familiarity with me.

Therefore, my attempt to pet her resulted in a bite and a permanent scar above my left eye. This was not the fault of the dog, but the dangers of a familial habituation of a particular animal for an entire breed or species. You may have a positive habituation based on interaction with one animal; however, the history of that habituation is not carried over to other animals of the same breed or species. Therefore, we must be aware of how habituation forms a synecdoche in our experience. Each animal is unique and the individual that forms the part of our experience will not be the same when we meet other animals from the same species. Synecdoche is how our consciousness constitutes meanings in experience, but experience cannot be reduced to similar forms of habituated synecdoche.

Individuation: Derrida, and Violence

The second unusual feature that helps in the establishment of the particular animal synecdoche for the zoo is the individuation, personification, and anthropomorphism encouraged by the zoo. The animals that stood as representations for a zoo visit tend to be larger animals (also known as megafauna) that have characteristics that make them easier to differentiate than smaller animals. For example, in a 2004 survey I helped conduct of 171 children at Brookfield Zoo about their favorite animals, the six most popular animals were: dolphin, giraffe, tiger, penguin, zebra, and lion. These animals together accounted for 60.8 percent of the sample.[1] These animals tend to be larger vertebrates that are characterized by a distinguishing feature (such as the zebras' stripes or the giraffes' neck), or a high level of activity by the animal in the zoo (such as the dolphins and penguins), or a sense of danger (such as the lion). The other feature of these animals is that they are the animals that the zoo tends to name. Some zoos even involve the public in contests to name newborn pandas, lions, dolphins, and other large vertebrates with one of the aforementioned characteristics; but there are never contests to name the new baby naked mole rat, tarantula, gazelle, or milk snake. Despite these elaborate public relations campaigns around the naming of zoo animals, sometimes these names are not carried over by the zookeepers and/or researchers in their behind the scenes work.[2]

In a paper he delivered at Duquesne University in February 2007, Leonard Lawlor stated that Derrida viewed the practice of giving individual names as an act of violence.[3] This goes back to biblical times when Adam named the animals as a way of having power over them. Therefore, Derrida refused to provide an individual name for his own cat that he writes about spying on him in the bathroom.[4] Derrida's analysis in this section is sharp funny, and very perceptive. He asks the question what does the animal see when it gazes at me? In particular, what does the animal see when he gazes at my naked body? Should I feel shame before the gaze of the animal in my nudity? There is acumen in Derrida's analysis and he makes a very interesting turn of events by focusing on the animal's perception of the human. Derrida shifts the analysis from a gaze at the animal to a reflection on the animal's gaze and its impact on the human body's manifold way of being in the world. It also conjures notions of Eden and Genesis.

The concern Derrida has with the notion of the animal is ultimately to decrease animal (and all) suffering in the world.[5] Yet Derrida is neither satisfied with a reduction to sameness that treats the animal with kindness by stating that the animal is a species just like us humans and requiring of the same "compassion" as with which we would treat other people. The problem with this response is that there is a reduction of all animals into one "continuous species."[6] However, the other response is one that maximizes difference and treats animals as so alien from the human that they are reduced to "mere machines." This was the analysis of modern philosophers from Descartes on that allowed them to turn a blind eye to the problem of animal suffering by denying that animals were living, and instead, were machines incapable of suffering. The capacity for suffering, as a mode of living, was reserved only for humans; animals were not considered to be alive in a similar fashion to us.

Derrida makes two objections in terms of how we speak of the animal. The first objection is to the individuation in naming. For Derrida the act of naming an animal harkens back to Genesis and was the way for Adam and his descendants to hold power over animals. By giving them names there is an act of power that separates man from animals. The taxonomic is man's domain; we alone give names and with this power of naming we stand above and separate from the animals. With the word, we pull ourselves out of the world next to animals and they become lesser than us. In this second tale of Genesis that grants Adam the power to name the animals, humans transcend the world and animals are stuck in the world like all the rest of the objects that we name.

Derrida's second objection in how we speak of the animal is in the use of the broad generality of the word "animal" that glosses over the difference and uniqueness of the different species and individual animals. Despite our references to the "rational animal," "narrative animal," or "playful animal," animal is also a separation of all the creatures from human beings.

> There is no animal in the general singular, separated from man by a single indivisible limit. We have to envisage the existence of "living creatures" whose plurality cannot be assembled within the single figure of an animality that is simply opposed to humanity. This does not of course mean ignoring or effacing everything that separates humankind from the other animals, creating a single large set, a single great, fundamentally homogeneous and continuous family tree going from the *animot* to the homo (faber, sapiens, or whatever else). That would be an *asinanity*, even more so to suspect anyone here of

> doing just that. I won't therefore devote another second to the double stupidity of that suspicion, even if, alas, it is quite widespread. I repeat that it is rather a matter of taking into account a multiplicity of heterogeneous structures and limits. Among nonhumans and separate from nonhumans there is an immense multiplicity of other living things that cannot in any way be homogenized, except by means of violence and willful ignorance, within the category of what is called the animal or animality in general. From the outset there are animals and, let's say, *l'animot*. The confusion of all nonhuman living creatures within the general and common category of the animal is not simply a sin against rigorous thinking, vigilance, lucidity, or empirical authority; it is also a crime. Not a crime against animality precisely, but a crime of the first order against the animals, against animals. Do we agree to presume that every murder, every transgression of the commandment "Thou shalt not kill" concerns only man (a question to come) and that in sum there are only crimes "against humanity?"[7]

Derrida understood the paradoxical synecdoche that is our relationship to animals. His concern with the violence of the individual particular and the violence of the same show the trouble of talking of animals in our taxonomic scientific language. The animal is a particular animal, it is this species of animal, and it is all animals.

I want to say that I share Derrida's concerns, but want to take them up in the context of our discussion of synecdoche operating in the zoo. Let me say I agree with his modernist critique of the reductive use of the word "animal" that obfuscates the animal into a broad general category of sameness. Despite the difficulty of knowing where the line between human and "animal" begins, we are different; to put us all under the label animal is an act of violence that reduces the heterogeneous into the homogenous. But just to use the label animal for all "nonhuman living creatures" is also a violent act of reducing multiplicity to singularity. This is a crime because it allows for a split between human and animals in a hierarchical relationship of power where the one may kill the other. With the reduction of all the different animals to the category of animal, separate from the human, Bentham's question, "Do the animals suffer?" becomes irrelevant. Suffering is severed from the individual animal because they are reduced to an abstraction of the general category animal.

Now, where I want to respond or leave Derrida for pragmatic reasons, is in his resistance to naming. While he may be right in the biblical exegesis, I object by stating that even a popular name, such as cat, is a name and this points out the futility and impossibility of avoiding naming. Derrida's ethical concern is with the problematic separation and wielding of power that comes in naming. So Derrida would object to zoos and even individuals naming individual pets in their household. (Does he also object to the human parent naming their human child?) This is a wielding of power and a radical separation of human from animal that would allow for their indiscriminate killing and suffering.

But, I think there is a different danger in relation to naming and animals for which Derrida is not accounting. The avoidance of a name is often used as a way to reify the animal and subsequently make the torture or slaughter of animals more bearable. For example, the horrendous practices of factory farming are grounded in the fact that these acts are not being conducted upon individual animals, but instead on a universal object.[8] So while I agree with zoos who want to avoid anthropomorphism by placing individual names on animals, there is a problem that the failure to place a name on an animal contributes to the animal's becoming a mere object. In the reification, the wielding of violent power by humans over the alien other animal also becomes possible. It is the name that calls forth the animal from the world of objects and to the active, living world capable of suffering.

The naming of an animal has a major impact on a child's interaction with a zoo animal. Naming allows for individuation where the child can pick out the individual animal from other members of its species and animals in its exhibit. Naming also is more conducive to establishing a connection between the animal and child because of its role in personification. By personification, I mean that the child can more readily connect particular behavioral characteristics with the animal that has that exact name. For the child, this differentiated named animal is then connected with particular behaviors, such as Olga the fun-loving walrus that blows water on the children, or Samson the majestic silverback whose power is evident in the strength with which he beats his chest.

Zoo Memories

The final aspect that allows an individual animal to stand as a synecdoche for a zoo is often times a memorable experience. When interviewing adults about their memorable experiences with animals and nature, I was surprised by the number that recounted stories with animals in the zoo.[9] There is a narrative structure to these memorable encounters. The memorable experiences hold a phenomenological structure with at least one of the following: *close contact*, *witnessing a peculiar activity*, or a *mystical description of harmony with the animal*. The eidetic narrative is exemplified in, "I remember when Babe the ox licked me with her big, strong sandpaper tongue," "I remember seeing the camels mating—it was very disgusting," or "I remember how much I used to love going to Wolf Woods in order to sit and watch the wolves. I could have been there all day. For some reason I felt like I was part of their pack." These are three typical narratives that provide a template for the memorable experience that allows for an individual animal to stand in representation for the entire zoo in the mind of a child. These narrative structures also tend to be preconditions for establishing some sort of bond with an animal. Zoos understand that close contact, unusual activity, and mystical connections bond the visitor to their animals. That is why, despite our litigious society, there has been an increase in exhibits such as Pittsburgh Zoo's Ray Petting Pond, which allows children under some supervision to pet miniature rays as they swim by. And zoo signage tends to emphasize the unusual and mythical elements of an animal in order to help establish these bonds and intentionally tap into this narrative structure.

Yet, this synecdoche is paradoxical. For most children understand at an early age that the animals in the zoo are confined, despite the lack of bars. By the age of ten, many children I interviewed had already developed a sort of *ecological awareness* that allowed them to understand the concept of a natural environment and that the zoo animals (despite all the revolutionary innovations in design) were being seen "outside of nature." When I asked children why a zoo animal might be unhappy, these children with a sense of ecological awareness would say, "Because he is confined," or, "Because he is not in his natural habitat." These responses are more common when the "unhappy animal" discussed is a large animal—such as an elephant or a polar bear. Observing a giant bear pace, or an elephant display an obsessive-compulsive, repetitive behavior can be upsetting for the child. Yet, the ecological awareness is an understanding that the zoo animal's status as zoo animal is a complicated one full of paradoxes. The zoo is a source of individual confinement for particular animals, but is also a place that "cares" for individual animals as demonstrated by the tendency for zoo animals to outlive their wild cousins. The zoo as research facility observes individual animals in the zoo so that this knowledge can then provide "care" for an entire species.[10] Oftentimes we see the formation of early axiological thinking quite spontaneously. Some children offer utilitarian explanations that the zoo animals must sacrifice their freedom to save the species. Other children follow a more deontological explanation in that a zoo fulfills a duty that we humans have to try to repair the damage that we inflict on the "web of life" that surrounds us every day.

Zoos have done a good job telling their narrative. In the chapter after next I will discuss the zoo's self-construction as "conservation center." The conservation narrative is told in signs telling zoo visitors of environmental research being supported by the zoo around the world, participation in breeding programs,[11] attempts to save endangered species, and veterinary work done to save individual animals in the zoo and larger community. The success of the conservation zoo narrative is evident in the large number of children who express interest in becoming zookeepers or zoo veterinarians. Yet, the awareness that these animals are not free does not disappear. So, the individual animal is a synecdoche for a zoo and so is the paradox of zoo as place of care and confinement. A zoo's operation in a child's consciousness is an example of the complexity of imagination and the simple tendency for a youthful mind to hold two seemingly contradictory features simultaneously.

In phenomenological terms, this paradox of zoo animals as both confined and cared for tends to operate as a background lived-experience for the child. This is not something that a child often actively struggles with, yet it is still, as Husserl pointed out, an intentional experience. As Husserl said, "For even background lived-experiences are intentional. A universal life that is a life of consciousness through and through encompasses the specific acts, those that are specifically ego-consciousness, as well as the background consciousness."[12] The child's active imaginings of the zoo involves a foreground of animals and a background of paradoxical narratives. Like all intentional acts, features of the background have the ability to be foreground as the child's focus comes to an act such as Ziggy the elephant falling into his moat at Brookfield Zoo, a sign at the Bronx Zoo explaining the conservation activities of the zoo, or a poignant question from an adult about the animals' confinement and contentment. For zoos to successfully play the role of "model citizen" they must be willing to openly discuss these paradoxes with children rather than shy away from them.

THE PARADOX OF WILDERNESS

Zoos are often thought of in a dialectical tension with the term wilderness. To designate an animal as a "zoo animal" is to point out that it is not (or no longer is) wild. Yet, there are two important paradoxical relations with the notion of the wilderness that need to be explored because they have an impact on how we are to understand the role of zoos in contemporary society. These are paradoxes of definition and deficit.

Paradox of Defining Wilderness

When my wife and I moved into our new apartment we were surprised to find two domesticated rabbits hopping around in our open backyard. These two white and black rabbits were just loose. You could get close enough to toss them some food, but they would dart away as soon as you were close enough in proximity to reach them. At the end of our first week we met one of our neighbors with his two daughters ages four and six. The man asked if his daughters could play in our yard with the rabbits. After introducing myself to the man and allowing his daughters to play I asked him who owned the rabbits. The man responded—"Nobody, they are wild." Now having had a Dutch rabbit in my childhood I immediately recognized the white and black patched rabbits as being a domestic variety. However, my neighbor responded with the information that the rabbits had been outside for the past year and nobody owned them—so he and his daughters believed they were "wild."

What do we mean when we say an animal is wild? One of the arguments against zoos is that they keep animals "out of the wild." Wilderness is simply defined as that area outside of the home where animals live. It is very curious that wild can refer to squirrels in the city or grizzly bears in Denali Forest. For an animal to be wild, it needs to construct its own home independent of human interference. As the human population expands we overpopulate the earth beyond a sustainable rate we now find humans in every corner of the planet. Where is there a space that we do not touch and interfere? Even in the most expansive jungles and forests our carbon footprint has created damage and leaves a trace of humanity. Wilderness too is now a construction. We place limits and control the flow of animals. We cause harm and commit mass deforestation that endangers animals by destroying their habitat. Our direct actions have caused the extinction of countless species and endangered countless more. No animal exists anywhere on this planet without the encroachment of humanity on their habitat. Even if no human ever steps foot on an island populated by animals we impact those creatures through things like increased temperatures through climate change, raising water levels around the globe, and our pollution will travel miles across an ocean to places where no actual human footprint may be, but where our virtual footprint is actually quite large.

We hold a myth of wild frontier lands. But these places don't exist anymore. We have changed the world and with it the home for all the animals on the planet. Some are able to adapt and survive, but others find no solitude in the forest of the wilderness. As we develop further and further into areas that were once untouched and without humans we push out animals who once called that land home. More and more contact between "wild animals" and humans in urban areas are taking place. In expanding our habitat we have changed the area that once supported much wildlife. This means an epidemic level of potential extinction of animals and an irreversible damage to wilderness.

Ultimately the only solution will be massive rethinking of the way we humans live on this planet. We need to rethink our cities, deal with over population, and develop clean and sustainable patterns of living. What responsibility do we have to the animals whose lives we have so dramatically impacted? The solution is not as simple as opening up the zoo and letting all the Siberian tigers back into Russia, or releasing polar bears on Arctic ice flows. We have decimated the size of these endangered animals' habitats. We have put at risk their ability to find the food they need to support their social groups and future generations. The paradox of returning animals to the wild is that as zoos work on species release programs to bring back endangered species we humans continue to increase the pressure on multiple species by intensifying the size of our global footprint. The wild is gone. One needs no more than go hiking in your local woods, or national park and the evidence of people is everywhere. If this is where we are sending our endangered animals then we are sentencing them to death.

The problem of extinction is one of our own making and ultimately the solution will be one of our making as well. Like the mythical arc, contemporary zoos play an important role in ensuring the future of multiple species of endangered animals. The biggest "crime" of humanity is threatening the future of entire species of animals. We are not only responsible to the animals that exist as our temporal contemporaries, but we are also responsible for the death of future generations and ultimately entire species. With limited resources zoos fight to preserve and prevent extinction. In this role, they are worthy of the metaphorical title of arc. However, to make true progress "after Noah" we need to change the way we live on this planet, for we are impacting future generations of humans too.

What we need to do is start reversing our encroachment into the wild ecosystems of our animal neighbors. However, this can only be done with commitment on a global level to do the proper planning. We also need to develop a better knowledge and appreciation of animals and their habitats. Zoos provide research and training for zoologists working out in the field. The zoo also educates the public about the problems we are making for our animal friends. Returning to those education and conservation motivations for visiting zoos, we see that zoos can play an important role in teaching people about endangered species and enacting behavior changes that lead to a greater conservation ethic. Zoos provide us with contact and individuated connections to animals that we would have to travel around the world to potentially see. Zoos also provide an important education function for the general public by exposing a mass audience to lessons of particular threatened species and their habitats. They also provide us with experiences with endangered, or vulnerable animals that can stand for the species. By bonding with these animals in a zoo setting we can help create an awareness and desire to preserve other members of their species not in the zoo. In the next section, we will see why it is more important than ever for zoos to take their "educating the public" role seriously.

Nature Deficit Disorder

In his book *Last Child in the Woods*, Richard Louv[13] laments a "nature-deficit disorder" in our children. While interviewing thousands of children across the country, Louv was struck by a fourth-grader's response to his question, where does he like to play? The boy responded, "indoors, 'cause that's where all the electrical outlets are."[14] The jarring contrast of the play engaged in by the fourth-grader and Louv during his own childhood helped him coin the term "nature-deficit disorder." I agree with his hypothesis that children are outdoors less and have diminished contact with nature. I also agree with his second hypothesis that there is a correlation between the rise in ADHD (attention deficit hyperactivity disorder) and "nature-deficit disorder." As children have brought their play indoors at increasing levels, the rate of ADHD (and obesity) in the general population has gone up. One of the measured health benefits of being outdoors is that it is relaxing. It has a calming effect that has been proven to lower the blood pressure. Being outside helps in focusing, because it "pulls the plug" from the sensory overload of the modern everyday life of the modern cyber-child.[15]

The blame for "nature-deficit disorder" should not be idealized to rest solely on the youth of America. It's not only that "today's kids" are different in their preference to be inside, but that "today's parents" are spending a shockingly decreasing amount outside with their kids as well. The whole society has fallen into the trap of the indoors. One of the parents Louv interviewed summed it up best, "Something else was different when we were young: our parents were outdoors. I'm not saying they were joining health clubs and things of that sort, but they were out of the house, out on the porch, talking to neighbors. As far as physical fitness goes today's kids are the sorriest generation in the history of the United States."[16]

Families were outdoors. People knew the "wilderness" around their homes. This is not the case today. We spend most of our day staring at various electronic screens instead of living and breathing in "nature." Children don't climb trees, and play in the dirt. They don't recognize the local berries and mushrooms. Instead, we find ourselves increasingly indoors as our health worsens due to the negative effects of a sedentary lifestyle. The ignorance of fauna and flora in our society is increasing exponentially with each generation. Zoos can provide a buttress to nature deficit disorder by taking conservation education seriously.

The zoo visit is a paradoxical pseudo-outdoors, constructed-wilderness location. At the zoo, the children and adults can learn about the animals. They can learn about which human behaviors are most harmful to the various species on display and make changes. These individual animals serve as ambassadors for their representative species. Zoo animals inhabit a liminal space between wild and captive at the zoo. In our imaginations, these animals stand as representatives for their entire species. The paradox of this synecdoche is that the more the natural habitats of animals are threatened in the wilderness, the more we need zoos to provide the education and give us the connections to care for particular species and ultimately decrease the threat of global extinction. The zoo animals stand in for other animals, the zoo, and even nature itself. It is a place of motion and movement where kids can play. The next chapter will look at the zoo as a place of play and motion.

NOTES

1. Olin E. Myers, Carol D. Saunders, and Erik Garrett, "What do children think animals need? Developmental trends," *Environmental Education Research* 10, no. 4 (2004): 551, doi: 10.1080/1350462042000291056.

2. According to Mullan and Marvin (1999), there are three types of names used in zoos: scientific names (Latin-based label of genus and species, such as *Ursus arctos middendorffi*), popular names (what we commonly call a species, such as Kodiak bears), and individual names (the names used to refer to specific individual animals, such as Jim the Kodiak bear at Brookfield Zoo). The authors then document that during a couple of their interviews, zoo employees (keepers, directors, and researchers) may not use the same individual name for an animal that the public uses, such as at the Jardin des Plantes in Paris where animals were never named, or the Minnesota Zoo where researcher Sally Friedman referred to animals by common names followed by a number. See Robert Mullan, and Garry Marvin, Zoo culture , (Chicago: University of Illinois Press, 1999). www.waza.org/en/site/about-waza/who-we-are. This goes against my common observations with people in the zoo industry. I have seen cases where keepers/researchers have used variations such as diminutives of an individual name, or some sort of nickname such as "Lazy Jim," but have not personally come across examples of zoo employees referring to animals by their common name and a number in my local context.

3. This eventually became the book, Leonard Lawlor, *This is not sufficient: An essay on animality and human nature in Derrida* (New York: Columbia University Press, 2007).

4. Jacques Derrida, *The animal that therefore I am* (New York: Fordham University Press, 2008).

5. Lawlor, *This is not sufficient*, 1.

6. Ibid.

7. Jacques Derrida, "The animal that therefore I am (More to follow)," *Critical Inquiry* 28, no. 2 (2002): 415–16. www.jstor.org/stable/1344276.

8. Derrida shares the same concern for factory farming. Lawlor, *This is not sufficient*, 1.

9. Having conducted a majority of my interviews in a zoo setting obviously had a major impact on this reporting. However, even some informal interviews done outside of the zoo still recounted memorable experiences with zoo animals. I believe that the number of zoo animals described in "memorable childhood experiences with animals" will continue to increase as the wilderness continues to decrease and children's contact with animals tends to be increasingly limited to zoo spaces.

10. This was not the case until more recently. For example, Lincoln Park Zoo's Bushman was one of only five gorillas in the United States in 1930 because there had been little success in keeping gorillas alive for any period of time. Boston had a one-year-old male die in 1897; Ringling Brothers had two that lasted less than a year in 1924; the Bronx Zoo made two attempts to acquire a gorilla in 1911 and 1940 and neither survived a full year. Mark Rosenthal, Carol Tauber, and Edward Uhlir, *The ark in the park: The story of Lincoln Park Zoo* (Chicago: University of Illinois Press, 2003), 4–5.

11. Early zoos saw each other as competition. Now through various breeding programs and other cooperative efforts, zoos exist in a much more cooperative environment than they did fifty or even twenty-five years ago. Although resources, such as animals, tend to be shared much more liberally, there is still an element of control and pressure placed on third-world zoos that participate in these programs before they are allowed to display animals.

12. Edmund Husserl, *Analyses concerning passive and active synthesis: Lectures on transcendental logic*, trans. Anthony Steinbock (Boston, MA: Kluwer Publishers, 2001), 20.

13. Richard Louv, Last child in the woods (Chapel Hill, NC: Algonquin Books, 2006).

14. Ibid., 10.

15. Rush, who signed the American Declaration of Independence spoke of a connection with the earth as a "curative" treatment for the mentally ill, according to Louv, Last child , 45. Just as Aaron Katcher has documented a correlation between pet ownership and lowering blood pressure, Roger Ulrich of Texas A & M has sought to document how viewing and being in a natural landscape positively impacts health and specifically heart disease. See the following: Roger Ulrich, "View through a window may influence recovery from surgery," Science , 224

(1984): 420–21. http://www.ncbi.nlm.nih.gov/pubmed/6143402. Roger Ulrich, "Effects of health facility interior design on wellness: Theory and scientific research," Journal of Health Care Design 3 (1991): 97–109. www.ncbi.nlm.nih.gov/pubmed/10123973. Reprinted in Sara Maryberry, Innovations in healthcare design (New York: Van Nostrand Reinhold, 1995), 88–104. Roger Ulrich, Outi Lundén, and John Eltinge, "Effects of exposure to nature and abstract pictures on patients recovering from heart surgery," paper presented at the Thirty-Third Meetings of the Society for Psychophysiological Research, Rottach-Egern, Germany, 1993. Abstract published in *Psychophysiology* 30 (Supplement 1, 1993): 7. Roger Ulrich, Robert Simons, Barbara Losito, Evelyn Fiorito, Mark Miles, and Michael Zelson, "Stress recovery during exposure to natural and urban environments," Journal of Environmental Psychology 11 (1991): 201–30. http://dx.doi.org/10.1016/S0272-4944(05)80184-7.

16. Louv, Last child , 11.

Chapter Six

Playing at the Zoo and Kinesthesia

PLAYING WITH SEA LIONS

What we are shifting to in this section is a more cognizant attention to something historically becomes a part of consciousness. The generative question is: How does the zoo become a paradoxical synecdoche in the imagination of the child? Yes, the zoo tells its own narrative that the children pick up. However, since the explanation of zoos as conservation center is a more recent cultural phenomenon (to be explained in the next chapter) there must be some historical structure that the zoo holds in the child's psyche. My argument is that there is a generative structure of play and imitation that is rooted in this early imagining of the zoo as place. In this next section, we will examine play and imitation in the zoo setting, with an eye to genetic phenomenology, which Husserl describes thus:

> Another "constitutive" phenomenology, the phenomenology of genesis, follows the history, the necessary history of this objectivation and thereby the history of the object itself as the object of a possible knowledge. The primordial history of objects leads back to hyletic objects and to the immanent ones in general, that is, to the genesis of them in original time-consciousness. Contained within the universal genesis of a monad are the histories of the constitution of objects that are there for this monad, and within the universal eidetic phenomenology of genesis this very process is (explicated as) accomplished for all conceivable objects in relation to all conceivable monads. [1]

Let us begin our search for "the history of objectivation." This primordial investigation involves a stripping away toward that which has been forgotten. It is the elapsed possibility that lies coiled in the heart of all our intentional acts. Now to the genetic analysis of the zoo, what seeds of originary experience can we dig up?

I mentioned that my favorite animal at the zoo growing up was Olga the Walrus. For me, Olga was the zoo. She was my favorite animal that I always had to see and talk to. One of Olga's neighbors at the Brookfield Zoo was a family of sea lions. One of my favorite times of the year to visit the sea lions was during the "off season" from October through April. The zoo would be empty and I would have the underwater sea lion viewing to myself. When nobody else was around I would "play" with the sea lions. My mother used to dress me in bright fluorescent caps to keep warm. One day after taking off my hat and placing it next to the plexiglass of the sea lion underwater viewing, I noticed that one of the sea lions went right up to the hat and was very curious about it. She pressed her nose up against the glass and appeared to be pointing at the hat with her whiskers and nose as her body remained suspended. She was completely absorbed by this "curious object."

As I moved the hat the sea lion would follow it with her nose like a golden retriever waiting for a tennis ball. When I moved it suddenly the sea lion would swim suddenly. Eventually I would run across the viewing area holding out my hat as the sea lion swam with me in a game of tag. I would laugh and remember the feeling of joy and happiness as I played with the sea lion. I remember thinking that the sea lion recognized me and seemed to enjoy our game. It was my belief that when I walked into the exhibit the sea lion would become more active and swim to the spot where we could play. And I recall a complete contentment while engaged in this play. My mother would have to almost literally pull me from the exhibit.

Play has received rigorous philosophical examination over the past century, especially in the discourse of phenomenology, hermeneutics, and existentialism. The works of Gadamer and Sartre in particular have paid attention to play. However, the most thorough explication of play to come under a meaningful philosophical analysis may have been conducted by John Huizinga. His seminal *Homo Ludens: A Study of the Play Element in Culture* has a very interesting etymological engagement with the meaning of play in various cultures. For example, Huizinga points out that in Hebrew *sahaq* associates laughter with the concept of playing; while in Latin *ludus* serves as an umbrella term that covers all forms of play, except for the concept of joking covered under *jocus;* and Greek uses the suffix *–inda* to connote a specific act of play (such as *sphairinda*, meaning to play ball). The origin of our English "play" comes from the Anglo-Saxon *plega* that not only means play, but also has the additional meanings of rapid movement and playing a musical instrument.[2] Through his very phenomenological survey of play, Huizinga goes on to connect mind, fun, laughter, the irrational, opposition to seriousness, voluntary activity, its separation from the everyday, performance, ritual, relationship, mood, movement, and music with play. The richness of his investigation cannot fully be repeated in this section, but it is worth noting the tendency for play to be connected with children's games, yet the term has a nimbleness that allows it to be connected with the seriousness of the sporting games that adults play.

Childhood is defined by play. This capacity to play is not unique to humans. John Huizinga begins *Homo ludens* with the following statement:

> Play is older than culture, for culture, however inadequately defined, always presupposes human society, and animals have not waited for man to teach them their playing. We can safely assert, even, that human civilization has added no essential feature to the general idea of play. Animals play just like men. We have only to watch young dogs to see that all the essentials of human play are present in their merry gambols.[3]

The young of many species engage in various activities that would be considered play. The ability to play is seen as a sign of intelligence in a species. It is my contention that one of the originary experiences of the zoo is the structure of play. Since at least the time of zoo as zoological park, the zoo has been a place of leisure. The sociological analysis of the zoo by Mullan and Marvin (1999) was excellent for its historical description, but it was imbued with a seriousness that forgot that part of the child's original construction of a zoo as a place of fun and play. Those preschoolers who visit the zoo have not made the more complex cognitive connections of the paradox of individual confinement and species freedom through conservation practices. The genesis of zoo is in a context of leisure and playfulness.

Playing or Teasing?

What does it mean to play at the zoo? Or to play with an animal in the zoo? To answer this question I will look at three components of this apodictic structure of zoo constitution: *teasing*, *emotive intention*, and *imitation*. More will be said about the kinesthetic quality of play in the next section of this chapter. Let us begin with an important distinction of what play is not—teasing. One of the possible early connections that children can make with animals is through teasing. Teasing is the solipsistic interaction with the animal as object for personal enjoyment. It is a reified form of interaction, but it is a potential stance that a child could adopt toward an animal that would be a "bad faith" interaction that denies the animal any form of subjectivity, and instead treats the animal as any other object in the horizon. Teasing is a disconnected interaction, but outwardly may manifest behaviors similar to play.

Let us look at two examples. In the first instance a five-year-old girl roars with a lion sleeping across the moat. She turns to her mother and says, "Mommy, I am a lion," and proceeds to repeatedly roar as she crawls around hunting her brother. In the second case, a twelve-year-old boy sees the lion sleeping across the moat and repeatedly roars at the lion saying, "Wake up, lion." All other things being equal, the behavior, the roaring, is the same. So what makes the first case play and the second teasing? This is a tough one that employees at a zoo must constantly navigate if they want to truly pass along an ethic of caring for animals to the children that pass through their gates. The important distinction here is the attention to the preposition "at" versus "with." As an employee of a zoo, I encountered both of these scenarios on the same day. One was important to stop and the other was important to nurture and reinforce.

Having worked at a zoo for over five years I do not hold a romanticized view of the interaction between children and animals in the zoo setting. I have seen children tease lions, pull peacock feathers, and roughly handle guinea pigs. These behaviors are done without an empathic connection between the animal and the child. The animal is mere alien object and the activity that constitutes teasing (for example roaring) is done completely for the egocentric gratification of the child. Oddly enough I often observe an adult or older child engage in these exact same teasing activities prior to the child taking up the behavior.

The difference between the teasing roar "at" and the playful roar "with" hinges on the child's *emotive intentional constitution* as such. Playfulness requires a playful emotive intentional constitutive stance. With play, the child is absorbed by the animal. The animal becomes foregrounded, and in particular there is an openness to the animal as generating nexus of its own experience. The child roars to have an experience that is like the experience of the lion. There is a passive anticipation that is pregnant with the animal's potential for activity that is a guiding component structure of the child's play. Ultimately the child's desire is for the animal to become play partner as the sea lion did in my childhood. There is also a noetic playfulness on the part of the child to engage in play. Playfulness has a connection with friendliness, happiness, and other positive egoic stances. The child must be in the "mood" to play, meaning that the child must have an attitude in the proper frame of mind to engage in the play. A simplistic way of thinking of this is that playfulness would be a structural necessity to engage in play.

Emotive intentional constitution is an object absorption of an animal Other. This animal Other must be thought "with" as having an analogical capacity for playfulness. (More will be said about this in the section on kinesthesis to follow.) Finally, there must be a noetic attitude conducive to play, such as happiness. I observed others tap on the glass of the sea lion exhibit merely to force the sea lion to move. These people were like the twelve-year-old roaring at the lion and lacked the emotive intentional constitution necessary for a play event. The animal interaction was self-absorbed and displayed no concern for the animal.

Imitation

The final aspect of play that will be looked at is imitation. In analyzing the child's play with an animal in the zoo it becomes necessary to look at imitation. The roar of the five-year-old girl involves imitation in so far as she is trying to *become like* a lion in her activity. Even my childhood play with the sea lion at the zoo involved a projected imitation. The sea lion imitated my motion of moving my winter hat on the glass of the underwater viewing with her swimming, and my movement was a circular projection of the desire for movement and imitation onto the sea lion. Also, my movement could be said to be imitating the movement of prey, or a game of tag or chase that I would engage in with human children. For the sea lion the chasing of my hat was an imitation of hunting a fish in the wild. However, let us be careful here. It is important to distinguish that genetic phenomenology of play is not the same as the explanation of play given by geneticists, behavioralists, or others engaged in psychologistic reductions of play. The genetic phenomenological account of play involves a primordial account of the history of objectivation that lies at the heart of a forgotten universal structure. The phenomenological account is fundamentally involved with the constitution of conscious experience; but for the geneticist who believes he/she can explain play through genes, play is reduced to mere instinct and is separated out from experience. Paul Guillaume makes a phenomenological critique of psychologism in his book *Imitation in Children* when he says,

> In rejecting the lazy hypothesis of a special instinct to explain imitation, we must connect imitation with primordial forms of activity. The perception of example is efficient only because, according to the principle laid down by Delacroix, "Visual perception . . . affects and releases a function or a group of functions" (20). Constituted functions (prehension, locomotion, sensory adaptations, etc.) rather than elementary movements are stimulated by perception. Imitation culminates solely in a transference that occurs at times directly; at other times by means of a training of sorts, to which the structure of instincts and of society gives a general direction that is in no way accidental. Imitation is not determined by the object but rather by the act of a human model; and finally by the very movements that constitute the act.[4]

A common explanation of play is that it serves an evolutionary feature and that imitative play is a result of an instinct that is displayed in order for the species to survive. So, the lion cub's play involves an imitation of behaviors needed for hunting; likewise human play is an imitation of necessary cognitive, auditory, and bodily behaviors needed for adulthood. Imitation in such a theory of play constitutes a means to an end and is therefore dismissed. However, children's imitative play with animals of other species presents a problem for the reductive psychologistic account. What instinctual benefit is there in imitating an animal outside of one's own species? Gauillaume is right that this account is "lazy" in failing to fully account for imitation as a constitutive act and not mere affective behavior. The mimetic elements of play cannot be reduced to imitative instincts.

The little girl pretending to be a lion is engaged in act of imitation that is an activity engaged in for its own sake. It is an intentional act. The girl sees the lion as an object of fascination that enters the foreground of her experience. She is in the right "playful" state of emotive intentional constitution. She roars with the lion. Her imitation is a desire to become "like" the lion. Her play is a genuine desire to interact with the lion as she recognizes the animal Other with a similar interiority possessing the same ability to engage in play as well.[5] There is a desire, or will, to play. Play's mimetic dimension is part of the originary constitution of the zoo animal experience for the child. The reason for a zoo's success is it taps into an early moment of playful genesis in which, as children, we desire to become like the animal Other.

Doing a genetic phenomenology requires a peeling away of the onion layer of experience to discover the forgotten hyletic core. So, why does the child imitate the animal in play? The answer is that the child's wanting to play is a form of communication. Play is a richly expressive act that contains a lot of important information about the child, animal, and the world around them. I mean it quite literally that play is a gesture of expression in the world. The child desires to communicate with the animal and thus realizes the need to connect with the animal, or become the animal. This is why, very often the first gesture of play between child and animal at the zoo is often a vocalization of the voice of the animal by the child. The child desires to sound like the animal in order to communicate with the animal. This originary desire to communicate with the animal becomes sedimented beneath various layers of experience as the child matures and loses the openness of the playful stance, and relates to the animal with various cognitive rememberings as to previous animal encounters. This desire to speak to animals is deeply seeded in us humans as is evident by the large number of different cultural myths that involve humans literally speaking to/with animals. The Dr. Doolittle desire to speak with the animals in play becomes lost in layers of adult experiences that habituate the child about how to act with animals.

In this section, we have looked at how play is part of the genesis of the experience of the child-animal bond in the zoo. Despite the separation provided by the moats, glass, and other "barless" barriers, the zoo can be a site of play for a child and an animal. This is seen in the differentiation between teasing and play, an openness of emotive intentional constitution, and a mimetic desire to imitate the animal. This habitual imitation as hyletic core of play is communicative. However, it does not mean that it is merely auditory. Yes, the child's zoo animal play may initially become manifest in a child's imitative vocalizations of an animal. However, in a further attempt to peel away a genitive core to the playful experience we will look at the kinesthetic elements that are apparent with the child's originary constitution of the play event.

"LEAPING LEMURS: LET'S MOVE IT, MOVE IT"

One of my jobs at Brookfield Zoo was to participate on a research team that evaluated an exhibit that sought to help children "develop caring attitudes toward the natural world." At Brookfield Zoo in Chicago there is a very intense exhibit called the Hamill Family Play Zoo that seeks to provide children with very intimate encounters with animals and their habitats within the context of play.

One of the areas of the play zoo that I want to focus on was called the "Lemur Leap." In this section, ring-tailed lemurs were rotated on and off display in a glass environment with a simulated forest setting. On the other side of the glass, the children had access to a similar enclosure with logs and forest setting material, lemur and zookeeper costumes, as well as various other props. I spent many hours in this exhibit observing the behaviors on display by both the children and the animals. The most exciting moments for me were when the lemur and child seemed to form a bond and actually play together. It was an amazing sight watching a little six-year-old boy or girl sitting on a fake log in a lemur costume, crouching down like a lemur and staring eye to eye with a lemur. The lemur would get close up to the glass on their side of the log and intently stare at the child and then all of the sudden engage in a giant leap to another log, with an "inviting glance" to the child who would immediately follow the lemur with their own leap (sometimes with shrieks of glee) in order to reset the starting position of the game. You could see the child "becoming" the animal in the play. Afterward the child would describe vivid sensations of what it was like to be a lemur even down to the feelings in "my own tail."

I had always been intrigued about what was actually going on in the play of the child and animal. As an engaged observer in the exhibit, I "felt" the connection between the child and animal. It was interesting to see this rich play and know that something special was taking place. Yet as the social scientist observing the exhibit, and even more as a "trained philosopher," I knew I was supposed to be the detached observer, skeptical and doubting. In the mode of the scientist and philosopher, this child-animal bond remained untapped, but as a phenomenologist the richness of the experience was clear.

In this section, I will utilize Husserl's notion of kinesthesia in order to help explicate certain aspects of the child-animal bond. To do this, I will mainly rely on the recently translated and often ignored essay: "Foundational Investigations of the Phenomenological Origin of the Spatiality of Nature: The Originary Ark, the Earth, Does Not Move."[6] In this short essay Husserl fleshes out three important ideas tied to the notion of kinesthesia that will be useful for answering the question of what, if anything of importance, is going on in the play between children and animals? These three concepts will be the distinction between mobility and rest, analogous apperception, and the Earth as "originary ark."

First Movement: Mobility and Rest

Psychologists employing empirical methods (Husserl probably would have referred to them as engaged in psychologism) often have a hard time defining play.[7] Therefore, some have simply concluded that it should not be studied or should be examined only in the context of the larger behaviors that are shown in play. Since the psychologists have a hard time developing a clear definition of what play is; they believe that it should only be analyzed as clearly distinct behaviors such as foraging, fighting, and so forth (Does anybody else see the image of an ostrich with its head in the sand here?) Despite the difficulty of adult researchers to define play, it is an actual experience of both humans and animals and deserves a proper philosophical examination. It also is something that is intensely engaged in during childhood and some (such as Piaget) have argued that it has important developmental significance. Influenced by the phenomenological tradition, Piaget even had a taxonomy of play that distinguished practice play, symbolic play and games with rules.[8]

Play can take many different forms. It can be solitary or social. Both humans and some animals engage in play. One of the most common observations of children's play with animals is that it involves an intense focus on mobility and rest. In order to engage the animal in play the child will intentionally utilize patterns of movement. For example, one afternoon I came across a Chihuahua puppy. After a brief discussion with the owner of this three-month-old pup, I decided to play with the puppy. So, I got down on the ground with my forearms and knees on the ground, looked at the puppy, was very still, and then very quickly jerked my head from side to side while tapping the ground and moving to my right. While playing with the puppy I didn't think about doing any of these actions; I just thought, "I want to play with this puppy. Let's see if we can get her to jump around and chase me." The puppy instantly recognized what I was doing and leapt around me and started to lick my face while wagging her little tail as fast as a hummingbird.

Play for both the child and the animal involves a strong perception of the motion, weight, and position of the body, muscles, and so forth, and their movement. This is what kinesthesia is. For Husserl, the analysis of kinesthesia involves an inner and an outer dimension.[9] This is because the kinesthetically engaged body is both outwardly physical and inwardly oriented, living flesh. There are two different manifestations of the bodily—*Körper und Leib*. Anything that partakes of this life-world of Earth has a physical body or a

Körper. Every *Körper* contains its own horizon of potential mobility and rest. To be a physical body means to possess this outward manifestation of either motion or rest.

The *Leib* contains another horizon of mobility and rest. The *Leib* first appears as with inner horizons. (It is important to note that Husserl is not introducing a dualism here. The kinesthetic is only meant to be analyzed upon a return to the life-world after passing through all the other stages of the phenomenological method from *epochē* to an analysis of the noematic and noetic pole of intentionality. Kinesthesia is a unity in a very similar fashion to the way the distinct noematic-noetic poles of intentionality are unified.[10]) The lived body of the *Leib* for Husserl is carnal, lived, unified flesh that is positioned on the Earth.

> Consider my flesh. In primordial experience, the flesh has no moving away and no rest, only inner motion and inner rest, only motion and inner rest, unlike the outer bodies. In "I am going," in any "I am moving myself" kinesthetically, all the bodies do not "move themselves" and the whole earth-ground under me does not move. For what defines a bodily rest is that the aspects of the body flow kinesthetically "in a mobile way" or do not flow, according to whether or not I hold myself still, etc. I do not move away; I stand still or go; thus my flesh is the center and the bodies at rest and moving are around me, and I have a ground that does not move. My flesh has extension, etc., but no change and nonchange of location in the sense that an outer body is presented as in moving, coming closer or moving farther away, or as not in motion, near, far away. But the ground on which my flesh walks or does not walk is not experienced as a body to move away from or not *completely*.[11]

What Husserl is saying here is that the lived body forms its own inner horizons of mobility and rest. From this inner stance the "lived flesh" experiences the world not as some absolute world-body in motion through the universe, but instead, the *Leib* is the center of its own horizons, and thus determines movement and rest from its own position. (Now you can see why Husserl felt this essay in a way overturned the Copernican system.)

So, back to my observations of the child and the lemur: From the standpoint of *Körper* analysis, there is a horizon of mobility and rest going on in their exchange. As an extended body myself, I see these two other physical bodies alternating between periods of mobility and rest. From my position and from their positions in relation to each other, there are both rest and movement oscillating back and forth in a frequent rate as child and lemur move , and away from each other.

What about the horizon of the *Leib* of child and lemur? For the child an important distinction needs to be made between playing at being a lemur, where a child will put on the costume and jump around with friends, in a sense completely oblivious to the lemur versus the richer playing *with* the lemur that I am interested in here. To play *with* the lemur the child must will it. There needs to be an inner movement of the child toward the lemur itself. This involves putting on the costume, approaching the exhibit, and looking for the lemur. The child will initially see the lemur as *Körper* in the tree, but in order to engage in play, the child must have an inner movement of hoping to play. There is the visible body of the child, the visible body of the lemur, and then there is this desire or will to play by the child. This desire to play involves an inner horizon of mobility and rest. The mobility is in what I call the child's *invitational gesture of play* that is the desired approach by the child, the getting into the role of being the lemur and the looking toward the bodily lemur with a desire to play fully saturating their being, and an intentional inner extending toward the lemur. All else in the room becomes irrelevant for the engrossed child.

The horizon of rest for the child involves a surrendering of roles in order to engage in the play. In order to play *with* the lemur, there must be recognition on the part of the child that the lemur, too, is a subject with its own potentialities that include the ability and desire to play. The child wants the lemur to enter into the play of its own "free will." For the child to play, the lemur itself must "want" to play. If the lemur was forced, or if it was only accidentally engaging in the outer bodily movements of alternate movements of gazing, leaping, and resting that constitute the play in this scene, then this would not be play for the child and it would not be fun. This recognition of the lemur as play subject wanting to engage in play is revealed by the internal and external rest of the child. The child has said that he/she wants to play with the lemur, put on the costume, and approached the exhibit. Yet, the child cannot immediately begin in play with the lemur. There must be a pause, an internal horizon of rest as the child "waits" for his/her lemur partner to signal recognition of the child's desire to play and an outer bodily display of "wanting" to play on the part of the lemur. For the play to begin, the child must follow the leaping lemur.

Second Movement: Analogous Apperception

It is important to note the experience of the engaged child is such that it is physically at rest looking for some bodily movement from the lemur, and at rest as the lived subject of child, inviting the lemur to engage in play and waiting for a response. For the child is anticipating that the lemur will play. Anticipation here implies the horizon of rest. However, it also implies a *responsive gesture of play* on the part of the lemur. From the child's position, the lemur is recognizing their invitation to play, has the potential and capacity to play, and then is intentionally signaling their responsive gesture that I define as a three-part gesture of external bodily motion of leaping, an internal sign that the lemur itself possess the potential for recognizing the child's invitation, and then "freely" choosing to respond to the child and beginning the play. That the lemur is its own internal *Leib* is central to the child's experiencing of play with the lemur. For the child the very foundation and potential to engage in play with the lemur is predicated on the fact that the lemur must possess a capacity and desire to play like me.

This is where Husserl's notion of *analogous apperception* comes into play in the analysis of kinesthesia. According to Husserl:

> The flesh of others are bodies at rest and in motion. But they are carnal bodies in the form of "I move," in which the "I" is "another I" for whom my flesh is a body, and for whom all outer bodies which are not carnal bodies for it are the same outer bodies as mine. But every carnal body as well, which is for me the flesh of someone else, is for all other egos identically the same body and the same flesh of the same ego. Likewise for every ego, my flesh is the same body and at the same time the same flesh for the same ego that I myself am for myself.[12]

So the transcendental move here for Husserl is that when I put myself in the position of the Other, I see that they, too, have an ego like me and live in a bodily world seeing various objects including me, which exist as objects. Yet, I know that I am not merely a *Körper* for myself or the other subject. I think that many critics have been dismissive of Husserl's notion of *analogous apperception* because they have failed to fully grasp its transcendental character. *Analogous apperception* is not merely seeing the other like me and having similar experiences. Instead, *analogous apperception* is a movement of shifting of one's nexus of horizons to that of the other and literally gazing out of the eyes of the Other. It is from this position that we then can imagine coming across the image of our own physical body. This image of ourselves combined with our previously bracketed experiences of self can help us understand that the world is full of other physical bodies that are more than merely physical, but are subjects like me.

This question of *analogous apperception* becomes really interesting when human relations with animals are considered. One of my annoyances with continental philosophy has been an ignoring of the question of the lived-experience of the animal Other. Instead, most of the philosophical literature about animals deals with whether or not they possess the right to be our food, be experimented on, and to what degree their consciousness is lesser than ours. However, there is a failure to try to understand the animal experience as it is lived by the animal. To truly understand animal experience, a sharp practice of analogous apperception is required. Yet, look how animals have been dealt with by Heidegger, for example: "The leap from living animals to humans that speak is as large if not larger than that from the lifeless stone to the living being."[13] From this leap of separation one could see how it would be easy to take another leap that Merleau-Ponty critiques Malebranche for making in his lecture on *Nature* : "It is in this way that Malebranche would not have beaten a stone as he beat his dog, saying that the dog didn't suffer."[14]

There is a problem with viewing the animal as radically Other, or pure physical body such as a stone. Part of the motivation for this view of animal as radically Other is that then it can be used to justify animal abuse. But is the animal radically Other? For Husserl the answer is no—I am able to think myself into the perspective of at least certain animals through analogous apperception. Let us examine how Husserl brought analogous apperception to bear upon a bird:

> Let us now assume that I am a bird and can fly—or assume that I watch the birds that also belong to the earth. To understand them is to put oneself in their place as flying. The bird sits on the branch or on the ground, then leaps into the air and flies upward: the bird is like me in its experiencing and in its action when it is upon the earth; it experiences the ground just as I do, it experiences different bodies, also other birds, the flesh of others, and carnal egos, etc. But the bird flies upward—that is like walking upside down, a kinesthesia whereby all courses of appearance, otherwise perceived as rest and motion of bodies undergo variation and in ways similar to walking.[15]

For Husserl, there are more similarities than differences when it comes to how the bird and I experience the world. There is not a large "leap" that can't be crossed between human and animal. Instead Husserl sees the human and animal as sharing certain aspects of experience that go with cohabiting the same world. When there is something as different as "flying" at work that I must imagine, then there is a second internal analogous apperception performed. Husserl externally thinks himself into the bird and then internally thinks himself into his originary mode of Being searching for similar experiences to understand the "unknown kinesthesia" of something like flying. So there is no radical alterity in the animal Other for Husserl, for a corresponding variation of the kinesthetic can be found.

I think of the usefulness of Husserl's account in bringing a new critique to the philosophy of mind issues brought up by Thomas Nagel in his famous essay, "What is it like to be a bat?" As mentioned earlier in chapter two of this book, I agree with Nagel's goal of wanting to counter the physicalists' accounts of mind, but I think Nagel failed to take into account the power and workings of analogous apperception. I may never exactly be able to experience what it is like to truly be a bat, but that doesn't mean I can't imagine what it is like to be a bat based upon my own experiences. We have a bundle of various kinesthesia. In our lives, we are aided with understanding our own experiences by cross-checking and comparing this new experience with older existing ones. After all, who I am is the unity of this entire "kinesthetic system." The bird or bat can be understood by substituting the missing experience with a similar one and by taking a look at the entire "kinesthetic system" through analogous apperception.[16]

Back to the lemur and the child—the child may not know what sounds a lemur makes and may immediately map on a monkey sound based upon the appearance and behavior of the lemur. Once the child spends time with lemurs, or is told that lemurs don't make monkey sounds, he or she then adjusts their corresponding analogous apperception as new information is added. So analogous apperception is never a finished and closed process, for I am not the Other qua Other. The child adjusts his or her perception of the animal in the physical world and restructures their analogous apperceptions of the unity of the Other's internal experience. This salient adjustment to the child's play is then made to remove the monkey sounds from the child's analogous apperceptions.

Likewise through analogous apperception I may link the bird or bat's experience of the kinesthesia of flying to "upside-down walking." Since my kinesthetic system is constantly changing, this means that I may be wrong or later discover a better analogous structure to understand what is going on in the experience of other animals.

Third Movement: Earth as Originary Ark

It seems like there is a danger of lapsing into relativism if we merely acknowledge the difference of perspectives of children, lemurs, birds, bats, and philosophers. So how is this phenomenological kinesthetic analysis not relativistic? How do we avoid ending up with the child in their own enclosed little cage and the lemur in his?

First of all, the lived subject is never merely reducible to any one particular kinesthesia. We are an open-ended bundle of kinesthesia. Secondly, and most important for Husserl, we all share the same earth. The earth is "a primordial ark"; it is the grounding point from which we all orient out together. As Husserl said, "There is only one humanity and one earth—all the fragments that are or have been separated from it belong to it. . . . [T]he earth is the ark which makes possible in the first place the sense of all motion and all rest as mode of one motion."[17]

We are all grounded on the earth. (By earth here, Husserl means any possible "primordial homeland" or earth-ground.) We live in the same world of objects and have the same bodily objects encountering our lived flesh. Husserl uses this metaphor of the ark for the world in contrast to the notion of a Nature of scientism. For Husserl, the "science of Nature" has introduced a dangerous infinity into the world where the flesh is only "accidentally particularized" that allows for the possibility of conceiving of a world without the living flesh. To quote Husserl, "in these sciences of the infinity of the totality of Nature, the mode of observation is usually the one in which carnal bodies are only accidentally particularized bodies which can therefore also conceivably be completely ignored so that a Nature without organisms, without animals and humans is possible."[18] For Husserl, the relativism of science and the relativism of solipsism are both overcome in the originary metaphor of the ark. The ark is that which is outside of all of us. It is an originary apodictic grounding shared by all living creatures on this planet. Therefore an ark is always a reminder of the more than one. The ark is also the bearer of life. It prevents us from being lost on the sea of scientism's infinity. The earth as ark is that starting point beneath our feet that all living creatures share. It is a useful metaphor because an ark is the ship that provides a place of sanctuary, living, and survival. Noah was instructed to build the ark in order to save himself and all the animals. Too bad humans seem to forget this sharedness quality of our ark. We allow ourselves to think that we are the only ones on this vessel and that we do not have to think about the Other's we hurt in our destruction of the earth. (That is why plans such as the Bush administration's justification of drilling of the Arctic incorrectly utilize the "earth as resource" metaphor to be used individually when, in truth, the earth is not something that can be relativized universally by objective science, or relativized individually as merely existing for the sole myopic purposes of one selfish group. When will they learn that drilling holes in our ark on stormy seas will cause us all to sink?)

For my example of the child playing with the lemur, there is an important element that needs to be examined. Employing the ark metaphor here allows us to look at the earth-grounding of both child and animal. First of all, there is an important set of similarities between the child and animal in terms of position. The shared location and proximity allow for this play. One of the assumptions of play is that the same world is operating about all those involved in the game. A refuge is created where the play takes place in extended space. However, in my zoo example, it is important to note that the play does take place in a zoo. What this means is that the freedom of the animal is impinged upon in order for the play to take place. That is why the types of play children engage in with animals in zoos, homes, farms, and the wild are completely different.

Ending Play

A very interesting analysis of the play of a little girl and wild animals is recounted by the philosopher Irigaray. She talked about how she loved to play "dress up" with the wild rabbits in her back yard. She also describes the experience of joy involved with having butterflies land on her. However, the part I would like to end with is her interpretation of her childhood play with birds.

> Birds are our friends. But also our guides, our scouts. Our angels in some respect. They accompany persons who are alone, comfort them restoring their health and their courage. Birds do more. Birds lead one's becoming. The bird's song heals many a useless world, it makes the breath virginal again and helps it rise. . . . The bird consoles, gives back to life, but not to inertia. The bird animates breath while safeguarding its materiality. It is more than overly logical speech, the pathway to restore but also transubstantiate the body, the flesh.[19]

For Irigaray animals were a source of a message of healing. They helped her cope in the harsh world. She did not believe in the domestication of pets. These animals were her friends that moved around the originary ark of her grounded-earth with her. Through analogous apperception she interpreted their angelic messages and they brought her comfort. Irigaray's development was tied to these playful encounters.

In this section, I have used Husserl's often-overlooked essay "Foundational Investigations of the Phenomenological Origin of the Spatiality of Nature: The Originary Ark, the Earth, Does Not Move," in order to provide a framework for doing a kinesthetic analysis of a child's play with animals. I have shown that it can be understood as possessing a mobile and rest horizon, involving analogical apperception, and finally taking place within the context of our shared ark, which all species have a claim to. Play and the child's relationship with animals are very important developmental elements that can be important for stabilizing our earth-ground. They deserve a place in proper philosophic consideration and it seems only natural that as experiences they be analyzed through a Husserlian phenomenology. The sheer vastness of the child-animal bond begs that it be examined. As our society grows and changes with the increasing mechanization of the future, we need to examine the meanings of our connections with the world around us.

> Then he looked into the air, questioning, for overhead he heard the sharp call of a bird. And behold! An eagle soared through the sky in wide circles and on him there hung a serpent, not like prey but like a friend: for she kept herself wound around his neck.
>
> "These are my animals," said Zarathustra and was happy in his heart. "The proudest animal under the sun and the wisest animal under the sun—they have gone out on a search. They want to determine whether Zarathustra is still alive. Verily, do I still live? . . . May my animals lead me!"[20]

NOTES

1. Husserl, *Analyses* concerning passive and active synthesis: Lectures on transcendental logic , trans. Anthony Steinbock (Boston, MA: Kluwer Academic Publishers, 2001), 634.
2. Johan Huizinga, *Homo ludens: A study of the play element in culture* (Boston, MA: Beacon Press, 1960), 29–48.
3. Ibid., 1.

4. Paul Guillaume, *Imitation in children* (Chicago, IL: University of Chicago Press, 1971), 123.

5. It is interesting that animals tend to react differently to human adults and children. Are the behaviors of children more acceptable because they possess a feature of childhood, helplessness, and youth that can be detected across species?

6. Edmund Husserl, "Foundational investigations of the phenomenological origin of the spatiality of nature: The originary ark, the Earth, does not move," in Maurice Merleau-Ponty, *Husserl at the limits of phenomenology: Including texts by Edmund Husserl* (Evanston, IL: Northwestern University Press, 2002), 117–32.

7. Thomas Power, *Play and exploration in children and animals* (Mahwah, NJ: Lawrence Erlbaum and Associates, 2000). In this piece Thomas Power does an extensive survey of the "porous" literature on the subjects of children and animals' play. He begins his book by claiming that maybe one reason play is ignored is that the psychologists have failed to produce a satisfactory behavioral or functional account of what play is.

8. Jean Piaget, *Play, dreams, and imagination in childhood* (New York: Norton, 1962).

9. Husserl, "Foundational investigations." This inner and outer distinction of horizons of rest and motion is important for Husserl's central argument in this essay, the result of which he believes overturns the traditional Copernican view of the Earth. For Husserl, in a sense, there is no inner movement for the Earth from its position because it has no "grounding" or zero position from which to make this so-called movement clear.

10. As noted in chapter one of this book, Husserl was vehemently opposed to any sort of dualism. An object can never be completely independent of our conscious grasping of that object; intentionality is always already there. Kinesthesia is distinct from a natural world analysis of movement because kinesthesia recognizes that movement itself must always involve the implication of intentionality. Hence we have the two poles of *Körper und Leib* that cannot be separate, for any moving flesh must be the flesh of some body.

11. Husserl, "Foundational investigations," 123.

12. Ibid.

13. Martin Heidegger, in Michael Haar, *The song of the Earth: Heidegger and the grounds of the history of being* (Bloomington: Indiana University Press, 1993), 29.

14. Maurice Merleau-Ponty, *Nature: Course notes from the College de France* (Evanston, IL: Northwestern University Press, 2003), 166.

15. Husserl, "Foundational investigations," 124.

16. Thomas Nagel, "What is it like to be a bat?" *The Philosophic Review* LXXXIII (1974): 435–50 www.jstor.org/discover/10.2307/.

17. Husserl, "Foundational investigations," 130.

18. Ibid., 128.

19. Luce Irigaray, "Animal compassion," in *Animal philosophy: Ethics and identity,* ed. Matthew Calarco and Peter Atterton, trans. Marilyn Gaddis Rose (London: Continuum, 2004), 195–201.

20. Friedrich Nietzsche, *Thus spoke Zarathustra*, trans. Walter Kaufmann (New York: Penguin, 1980).

Chapter Seven

Zoos' Troubled Origin

Toward a Genetic and Generative Phenomenology

AN EVOLUTIONARY MODEL OF ZOOS

Zoos have begun to conduct analysis of their meaning and potential impact in the larger society. However, despite a few rhetorical and philosophical examinations, zoos have remained outside of the gaze of the humanities. I believe that they have been ignored either because they are viewed as insignificant, or because of the reductive presuppositions about a zoo's moral fallibility that it is merely a place for keeping animals in confinement. As we engage the phenomenological method, one of the first steps is a bracketing of prejudices. In this chapter, we explore another more vertical way of exploring presuppositions by doing historical analysis of how they became embedded. Now I am not claiming that visitors hold these views of zoos embedded in historicity, but I believe that the presuppositions attached to the zoo as it operates in the culture can and often do impact the general experience of a zoo. Whether we are aware of it or not, the history shapes the ground from which we engage in experience.

In 1907, Carl Hagenbeck finished construction of a revolutionary zoo outside of Hamburg, Germany. Hagenbeck made use of the newly invented "gunite" or "shotcrete" that was a form of concrete that was strong enough to contain the most powerful animals, yet nimble enough to be manipulated into faux natural objects. It was Hagenbeck's belief that the animals should be seen in as natural an environment as possible. He found the public zoos of his time to be depressing places with their bars that conjured up connections to human prisons. Hagenbeck pioneered the use of this gunite concrete to build naturalistic-looking environments for the animals, and then utilized moats and ditches to provide a safe distance between the animals and the human visitors. The implementation of this pioneering concept one hundred years ago marks the birth of the modern zoo.

There are at least three distinct phases in the evolution of the zoo. In their 1993 executive summary entitled "The Role of the Zoos and Aquaria of the World in Global Conservation," the World Zoo Organization adopted the zoo evolutionary model that had been used by the Chicago Zoological Society, which had been prominently displayed at Brookfield Zoo and created by Brookfield's director, Dr. George Rabb. This is the model:[1]

The Menagerie

Zoos have been around for thousands of years. The early zoos were a collection of exotic animals for the rich and powerful. They were private collections that demonstrated the power of the elite. Animals were kept in cages and the more exotic, powerful, rare, and distant the source of the species, the better the collection. The animals were kept for the pure enjoyment of the collection's owner, although other members of high society would be allowed to visit the collection as well. Examples of menageries were kept by the Egyptians, Greeks, Babylonians, and even by the Aztecs.

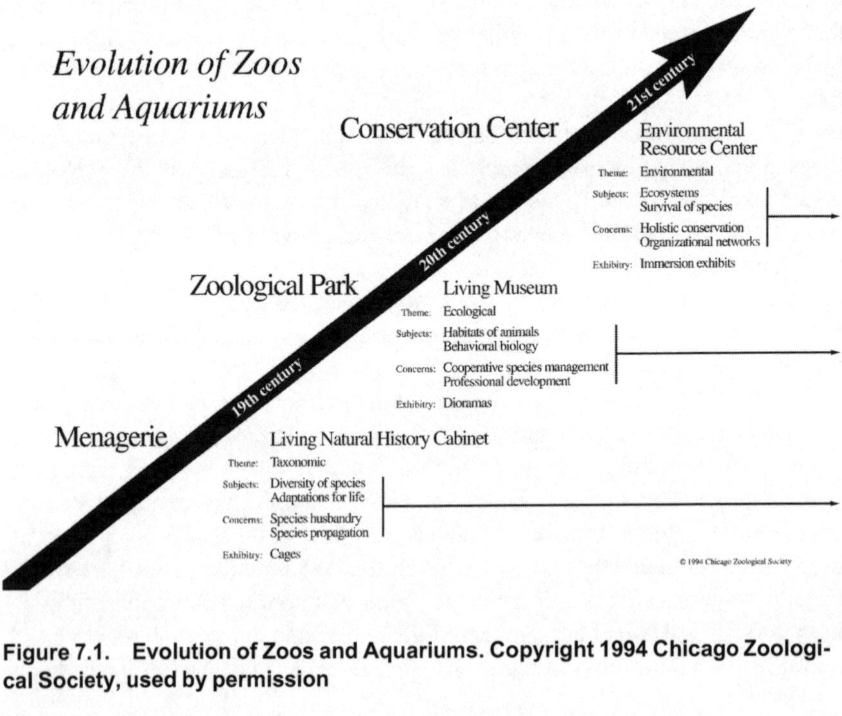

Figure 7.1. Evolution of Zoos and Aquariums. Copyright 1994 Chicago Zoological Society, used by permission

The earliest animal collections served either a religious or "princely" purpose. The animals in these menageries served a religious purpose in so far as they were kept because of their connection to a deity, or to serve as a source of animal sacrifice. The animals of the menagerie served a "princely" purpose when they were kept by the rich, elite, and powerful for their entertainment purposes, to assist in their "sport" of hunting, as gifts of honor, or to serve as exotic prey. For example, in Rome large collections of animals were kept for the "circus" of the amphitheater where the animals were pitted against each other (and often humans) for the entertainment of the public. Another example of the early princely connection of animal collections was Charlemagne's being sent a gift of an elephant and monkeys from the Caliph of Baghdad after his coronation. These animals became part of Charlemagne's already large collection of animals.

Early menageries were private collections that were kept with attention to gathering rare species. The "theme" of the menageries was "taxonomic" because the concern was with collecting as large a variety of species as possible. The successfulness and prestige of an individual's menagerie lay in being able to have as diverse a representation of species and from geographically as far away as possible.

By the end of the eighteenth and the start of the nineteenth century two major transitions took place in the evolution of the zoo: the establishment of public zoos and the switch to zoological parks. Since most of the early menageries were private, the public had limited opportunities to visit and view these collections. Quite possibly the first public zoo was found in Paris as the people of France struggled with what to do with the private collections of the elite after the French Revolution.

Louis XIV had established one of the most well-known collections of animals in the world. The design of his zoo represented the revolutionary shift in zoo architecture from pure menagerie to zoological garden that will be discussed later. The collection at Versailles was not well-maintained by either of his successors, so by the time of the French Revolution the menagerie was in disrepair. But it was still believed that these animals were kept in better condition than many of the average French citizens at the time of the Bastille uprising. Also, the revolutionaries identified with the animals kept in their cages as a symbol of the domination and oppression of the French king. Mullan and Marvin recount the record of the French historian Loisel's account of the revolutionaries confronting the king's director of the menagerie.

> In October 1789, a group marched out to the menagerie where they were confronted by the inspector in charge. The spokesman of the group shouted, "In the name of the people and in the name of nature summoned him to set free beings that had freely come out of the hands of the Creator and been improperly detained by the pride and the pomp of tyrants' (Loisel, 1912: vol. 2,

> 160). . . . The director, Laimant, pointed out that the first thing many of the animals would do if released would be to devour their liberators, and therefore, rather than freeing the animals himself, he would prefer to hand over the keys to the mob.[2]

I would have loved to have seen the expression on the faces of the mob as they thought about Laimant's point!

What to do with the royal menagerie of Versailles was quite a problem. During the Reign of Terror when many Parisians were starving, caring for the former king's exotic animals didn't seem like the right thing to do. Many resented the animals that were being fed while they starved. Others still felt a solidarity with the animals who had been trapped and imprisoned by the cruel king, as they felt all the French people had. Many of the animals ended up dying of starvation or were sent to slaughterhouses. However, under the guidance of the French naturalist Georges Louis Leclerc some of these animals ended up in the Jardin du Roi. (This former royal botanical garden became the modern Jardin des Plantes.) The menagerie of the Museum National in Paris was founded in 1793 with the explicit purpose of zoological study to examine the behavioral biology of the animals. This is credited with being "the first national menagerie in the world." It grew out of the remnants of the Versailles menagerie and other collections seized from the rich. The French collection became a source of national pride. Almost immediately after the success of this French public menagerie, other major European cities started establishing their own public zoos as a sign of prestige—not of the individual, but of the city and state.

The Zoological Park

The collection of Louis XIV at Versailles was famous for another reason. The design of the garden influenced the evolution of zoos as well.

> The designers of the earliest menageries or zoo buildings showed no real concern for the needs of the animals. Instead they were preoccupied with shaping that part of the building which surrounded the basic brick or iron enclosed area in which the animals lived. Menageries from the earliest times, such as that of Louis XIV at Versailles in 1662 or the imperial menagerie at Schönbrunn in 1752, were designed to please the visitors rather than to promote the welfare of the animals held in them.[3]

However, the expansive Versailles garden had a very interesting construction that would mark it as innovative for zoos. It was built in the gardens in an octagon shape. The king and his friends had a central house built in the garden. From there seven separate enclosures were "arranged" with various plants and animals. Each of the seven enclosures was separated from the garden house by an iron gate and each enclosure was shut off from the other ones with a solid wall. The areas were stocked with various plants and animals that could each be observed from the central balcony of the garden house. Despite the fact that the central concern was still for the humans' viewing pleasure and not the animals' welfare, the royal menagerie of Louis XIV was constructed as a royal garden that could be observed at all times. The important French philosopher Michel Foucault speculated that Bentham's panopticon was influenced by the royal menagerie garden at Versailles.

The important innovation at Versailles was the notion of the royal garden. The animals tended to be kept in more open-air confines. Versailles started the trend of larger, outdoor enclosures. The exhibition of the animals was switched from mere cages to a "caged garden," or park. Many early, public zoological societies picked up on this and called themselves "zoological parks." The notion of the zoological park was not only found in places like London's Zoological Garden of Regent's Park, but also in the Smithsonian National Zoological Park in Washington, D.C., and the Jardín (translates to garden) Zoológico de la Ciudad de Buenos Aires. It is in this shifting context from menagerie to park that Hagenbeck revolutionized the design of the zoo. The animals of the zoological park were moved from their traditional cages to dioramas. The visitors would walk past an exhibit and no longer see bars, but an open-air exhibit featuring the animal with faux rocks, trees, and various flora. There was an attempt for a more natural construction in the presentation of the display of the animal in their environment. The diorama was a construction of a quasi-natural-looking environment for the animal. However, the animal was still always accessible to the gaze of the public.

It is also during the evolution from the menagerie to the zoological park that the term zoo was made famous in an English folk song, and an explicit connection with scientists was maintained. The zoo was a place for education and culture and grew up alongside its more "refined" older siblings, the museum and art gallery. Mullan and Marvin have argued that since the zoo was defined with animals and their natural behavior, instead of interpretation and human artistic expression, or cultural achievement, zoos were connected with lower culture.

Although art galleries, museums, and zoological gardens shared many elements in their development, the two former institutions entered the realm of high culture, whereas zoological gardens became essentially popular and lower or common culture. Museums became associated with learning; they were places established by learned people, and this aura naturally conditioned responses to them.

Conservation Centers

The zoological garden continued as the dominant model for zoos until the middle of the twentieth century. During the past couple of decades, zoos have changed from zoological gardens into conservation centers. This shift was marked by innovations in the display of the animals and a cultural change in the role expectation of the zoo itself. The exhibition of animals in the modern conservation center is characterized by larger "immersion" exhibits. Some of the earliest examples of these were found at the Bronx Zoo in their Jungle World and Brookfield Zoo's Tropic World.

Tropic World, which first opened in 1982, was a massive 475- by 110-foot building that recreated a tropical rain forest indoors. The premise of the building was to change the animal display from the diorama to a new "immersion" exhibit where the animals were displayed in an environment that attempted to simulate the setting of the animal in its habitat in the wild. This included displaying plants and animals from the same area, a more expansive area for the animals, a decrease and camouflaging of exhibit signage, and even an attempt at imitation of events of nature like rain storms. Tropical World's most famous feature was its family of gorillas that were originally made popular by Samson the silverback and then by the controversial "rescue" by Binti Jua of a young boy mentioned in chapter 4.

The change from zoological park to conservation center is marked with zoos' increasing role in ensuring the survival of various species and conservation of nature. Zoos have moved beyond their own gates to see themselves as citizens with a role to play in promoting the conservation of biological life and the environment in the world as a whole. The modern zoo pursues conservation both as part of its mission as institution and as neighbor embracing its corporate responsibility to the community. This was the vision that William Conway, director emeritus of the Wildlife Conservation Society, saw in his 1999 keynote address to the World Zoo Organization:

> Thus the zoo's vision for the twenty-first century should be to become proactive wildlife conservation care-givers and intellectual resources; to step out beyond our fences by aiding parks and reserves; to sustain animals which have lost their habitats and conduct campaigns to restore them—and to provide from our collections as many key species as possible to be the stimulus and centerpieces of conservation efforts around the world.[4]

The twenty-first century vision of zoo as conservation center is exemplified by programs such as Brookfield Zoo's breeding program research in New Mexico to help repopulate the Przewalski's horse and the addax. Brookfield
o was active in Hawaii helping other zoos, the state of Hawaii,
Fish and Wildlife Service to protect endangered birds. They do
rch with South American penguins (having a geneticist on staff

year round), are working to repopulate the golden lion tamarins into Brazil's rain forest, and assess risk to wild dolphins from various forms of human contact. These are just a few examples of conservation programs that zoos are conducting beyond their fences to be involved in worldwide conservation.

The modern zoo now has four roles: model citizen, wildlife conservationist, agent for conservation, as well as mentor and trainer. The change from zoological park to conservation center is marked by a pedagogical paradigm shift as well. The old educational model of the zoological garden was the straightforward education of the "living museum." The pedagogical approach of the "living museum" provided education through treatment of the animals as artifacts to be collected and displayed. Now zoos see themselves as having a holistic, pedagogical function, instead of providing a mere encyclopedic function as a "living museum." The zoo as conservation center is a shift to an honest dealing with the zoo's embedded agency in the larger cultural context. For example, during the time period of zoos existing as zoological park, the phrase "endangered species" would have not been mentioned. A Siberian tiger would have been on display with signage that explained where they live in the world, their Latin name, and basic physiological features such as size and diet. However, in the conservation center the zoo's signage would not merely teach about these physiological features, but would explain that the Siberian tiger is an endangered species and what is being done to save this "vanishing animal." It is not enough for the zoo to provide entertainment and education; to survive in the next millennium, the zoo must recognize its existential responsibility to provide a broader web of care for those animals that serve as "ambassadors" in its setting. This involves a drastic shift from passive collector and displayer, to active animal advocate working to inspire visitors as well as to promote scientific research that protects and assists animals in a global situation that endangers many of the current species that exist on the planet.

ZOOS AND COLONIALISM

During the menagerie period it should come as no surprise that zoos displayed humans throughout the world. The racism and chauvinism of the colonial age reduced entire groups of people to "beasts" who were thought of as being animals. The racist reduction of the human to animal was most often done by those of European descent against Africans. What would surprise most people is how recently the last human was displayed in a zoo in America. In 1904 a missionary working in the Congo, Samuel Phillip Verner, brought back several Africans to be "displayed" during the St. Louis World Fair. The purpose was to display humans from around the world in all shades and sizes for a human evolution exhibit. Of course the racism that perpetuated across the field of anthropology held the Africans to be lesser humans and closer to apes then those "more evolved," white people of European descent. The biggest attraction of the fair was a pygmy named Ota Benga. Verner had purchased Benga's freedom in the Congo. Benga's teeth were sharpened as was the custom of his tribe and he was presented as a savage. During the World's Fair, he was mobbed to the point that the First Illinois Regiment National Guard had to secure his safety from the mobs at the fair.

Verner eventually travelled to New York and he and Ota Benga stayed at the Museum of Natural History. Since he had collected many items in Africa, Verner tried to sell several of these items. The young Bronx Zoo and its director William Hornaday were interested in "displaying Ota Benga" in their Monkey House. The zoo "purchased/compensated" Verner for Benga. And since slavery was illegal Benga was "employed" by the zoo to hang around the Monkey House and perform tricks. Benga even had a sign like the animals in the exhibit. Ota Benga's first day on exhibit was September 8, 1906. By September 16, more than 40,000 New Yorkers were flocking to the zoo just to see the poor pygmy. As an interesting side note, the Bronx Zoo also employed the scientist, geneticist, and eugenicist Madison Grant. The research Grant conducted while employed for the zoo helped justify many of the eugenics and anti-immigration laws that were passed in the United States at the start of the twentieth century. Eventually this research would find its way to Adolf Hitler and the Third Reich.[5]

Several black churches in New York complained about the poor treatment and display of Ota Benga, which *The New York Times* defended in an editorial, "It is absurd to make moan over the imagined humiliation and degradation Benga is suffering. The pygmies . . . are very low in the human scale."[6] The zoo decided to free Benga from the Monkey House and allowed him to roam the grounds. However, visitors would constantly accost him and there were a few violent skirmishes between Benga and visitors. At the end of 1906, the Bronx Zoo decided to no longer "employ" Benga. Unable to return to Africa due to a lack of money and the outbreak of World War I, Benga ended up moving to Lychburg, Virginia, to work in a tobacco factory. On March 20, 1916, at the age of 32, Ota Benga shot himself to death.[7]

This displaying of "lesser humans" was not unique to American Zoos. As a matter of fact, Hagenbeck, the designer responsible for the evolution of the modern natural immersion display and evolution of the zoo out of the age of the menagerie, "employed" Laplanders to stand on display with reindeer, and Africans to "care for" a variety of monkeys and apes. The colonial history of the zoo and its racist display of "lesser humans" is not known by most of the ever, this is an important history that zoos must come to terms l of hiding this past they need to account for these past practices e that no remnants of these horrific practices have in any way been carried over to the contemporary manifestation of the zoo. If the zoo is to fulfill its role as educator, then they should also educate us on the past crimes carried on in the name of science on their grounds.

After reading the horrific story of Benga, I couldn't help but think about how awful it must have been for Benga to be hit, harassed, prodded, and constantly provoked by visitors to the Bronx Zoo. As I imagined his torture, I envisioned families pounding on glass exhibits in our local Monkey House right on top of signs warning them not to stress out the animals by tapping on the glass. The passive zoo visit, far from increasing empathy and connectivity, could serve to breed feelings of superiority, elitism, speciesism, racism, and disconnect. That is why the story of the treatment of people, like Ota Benga, at zoos needs to be more widely known in both zoo crowds and the general public.

BIOPHILIA

We now move from the genetic phenomenological origin of the meaning of zoos and how they have evolved to become part of our historical consciousness, to investigate the originary ground for having zoos in the first place. What is the justification that has allowed zoos to come into existence? Now here I am not doing a natural world description of why we visit zoos (see chapter three), nor am I talking about their historical origin as examined above. Instead, the question is why bother having any zoos at all?

I hope the earlier phenomenological analysis in this book has shown you how much we desire to seek out other living things and love to be around them. We can spend hours sitting around and petting a dog, while even the most enthusiastic car owner would never touch their car in such an engaged manner. Our lives are filled with pictures of animals. We bring them into our homes, we tell stories about them, we seek them in the wild, and we visit them in the zoo. Yet we do this with plants, too. Across from me sits a pot with a yucca plant sitting in it. We seek out various living things and populate our life with them. We engage in this behavior at an early age. Children are just fascinated by all forms of living creatures and plants.

One of the popular explanations for this phenomenon of seeking living things comes from the famed evolutionary biologist and entomologist E. O. Wilson. "Biophilia" is a term coined by E. O. Wilson and it means, "the innate tendency to focus on life and lifelike processes."[8] He also explains it as follows:

> Given the means and sufficient leisure, a large portion of the populace backpacks, hunts, fishes, birdwatches, and gardens. In the United States and Canada more people visit zoos and aquariums than attend all professional athletic events combined. They crowd the national parks to view natural landscapes,

> looking from the tops of prominences out across rugged terrain for a glimpse of tumbling water and animals living free. They travel long distances to stroll along the seashore, for reasons they can't put into words. These are examples of what I have called *biophilia*, the connections that human beings subconsciously seek with the rest of life. To biophilia can be added the idea of wilderness, all the land and communities of plants and animals still unsullied by human occupation. Into wilderness people travel in search of new life and wonder, and humanized and made physically secure. Wilderness settles peace on the soul because it needs no help; it is beyond human contrivance. Wilderness is a metaphor of unlimited opportunity, rising from the tribal memory of a time when humanity spread across the world valley to valley, island to island, godstruck, firm in the belief that virgin land went on forever past the horizon.[9]

Wilson dreamed of turning "speculative philosophy" into a science.[10] Let us now move the opposite way and view Wilson's science as philosophy. As an evolutionary biologist, genetic determinism shaped a lot of his thinking. It is amazing to see Wilson come up with the concept of biophilia and hypothesize about its possible genetic origin. By his own account, Wilson's early childhood involved being around, recognizing, and interacting with various forms of living plants and animals. From the definition of biophilia one can deduce that those who were the most sensitive to other animals were the most likely to survive and pass on their genes. Yet by his own account of his childhood, Wilson's connection to nature was not the norm, but an aberration in comparison with his peers.[11] This genetic determinism caused a reviewer of *The Biophilia Hypothesis* in *Science* to remark "Wilson and his colleagues have identified yet another human behavior they suspect is governed by genes—nature appreciation."[12]

Biophilia is a belief that we have an innate, or at least a natural tendency, to want to affiliate with life. Investigations of the child-animal bond show that this occurs really early on. However, from Augustine to Descartes, philosophical conceptualizations of the human have sought to separate us from the world of the animal and nature. I don't want to discount important philosophical notions of rationality, but this strategy has had the consequence of painting an incomplete picture of the human condition. If we emphasize human rationality, individualism, and technological achievement, we must be careful to not forget that we are also part of the world as well. Psychological orientations of self, beginning with Freud, have also commonly sought to describe childhood development as closed, selfish, and merely inward. The initial "mytho-poetic" descriptions of the individual thrust into the world seem to at best be an exaggeration, and at worst be completely wrong. Our original orientations are outward and boundless—"oceanic." We literally are born one with the living world.

Let us interrogate the philosophy of Wilson's science. In the collection of essays *The Biophilia Hypothesis*, Wilson restates biophilia as follows: "Biophilia, if it exists, and I believe it exists, is the innately emotional affiliation of human beings to other living organisms. Innate means hereditary and hence part of ultimate human nature. . . . The biophilia hypothesis goes on to hold that the multiple strands of emotional response are woven into symbols composing a large part of culture."[13]

Wilson then goes on to give a snake and the human connection to it as the originary exemplar of biophilia. His argument basically is that snakes are poisonous and dangerous. Humans and other primates have a natural fear of snakes that manifests in a fascination of snakes and attention to snake-like objects as well as a specialized vocal communication of warning about snakes. This is due to early relatives who observed the danger of snakes, developed a healthy fear of snakes that resulted in behaviors that led to their survival through natural selection. In the transition from hunter-gatherers to modern civilization, we have less contact with snakes and consequently retain this fear in the realm of the symbolic. Wilson then points out that snakes are a subject of dreams in a wide variety of cultures.[14]

Let us first look at Wilson's use of an aversion towards snakes as his proof of the innate characteristics of biophilia. First of all, this seems like a poor example of biophilia. How can Wilson define biophilia as an "affiliation" toward life and living creatures and then provide his central example being a case where humans have an "aversion" to life. Aversion and affiliation seem to be opposites. The fear of snakes is not biophilia. Let us carry his example through. If I was afraid of snakes and saw them as a threat to my survival, then I would do my very best to kill these snakes, or at the very least, drive them out of my area as St. Patrick did in Ireland.

If Wilson wants to provide an exemplar of biophilia then he should provide something like a child's fascination with the movement of bugs, or contentment in petting their greyhound. However, Wilson can't provide a hereditary example of biophilia because these "true" biophilic examples involve emotions and aesthetics. We like animals and nature because it just "feels good." So why did Wilson grab, collect, and study snakes in the wild as a boy? Was it due to some innate gene that caused a fascination with snakes in him? It would seem that the species Homo sapiens wouldn't have gotten very far if lots of other little boys had genes that caused them to grab venomous water moccasins. Unfortunately, I agree with Holmes Rolston III's critique that Wilson mixes his brilliant observation of the human tendency toward affiliation with animals and nature with his earlier "selfish gene theory" (E. O. Wilson, *Sociobiology* [Boston: Belknap Press of Harvard University Press, 1975a]).[15]

So, I am agreeing with Wilson's notion that biophilia is an affinity with nature and his conclusion that this means that we need to do everything possible to secure the biodiversity of the planet because we don't know the future developmental consequences on humans of the massive loss of life.

However, Wilson's argument that biophilia is the result of an innate gene is as much hogwash as the recent arguments that there is a gene for love, the belief in God, or certain aesthetic tastes. Genes are a physical structure, biophilia is an emotional and aesthetic state. As a good phenomenologist, I acknowledge that bodily genes can influence our mental states and vice versa, but I cannot reduce a state of the human mind to a simplistic bodily explanation. To understand the origin of biophilia, a richer understanding of its symbolic nature and the various sedimented suppositions will need a rich genetic phenomenological explication.

LEBENSNEID

Nietzsche provides a strong critique of a certain type of envy that he called *lebensneid*, or literally "envy of life." Lebensneid is a characteristic of *ressentiment.* In *Nietzsche and Philosophy*, Gilles Deleuze defines ressentiment as when a "reaction ceases to be acted in order to become something felt."[16] Deleuze goes on to describe three characteristics of ressentiment: an "inability to admire, respect or love," "passivity," and "the imputation of wrongs, the distribution of responsibilities, perpetual accusation."[17] Ressentiment is the slave's spirit of revenge. So what does ressentiment have to do with lebensneid?

The connection is twofold: the first, definitional; and the second is metaphorical. Erich Fromm provided the definitional connection between ressentiment and lebensneid in his essay "Selfishness and Self-Love." After discussing frustration in the Oedipal complex, Fromm talked about the many forms that the child's reaction to this frustration may take.

> There are many forms in which the fight for freedom is fought and many ways in which the defeat is disguised. The child may be ready to internalize the external authority and be "good," it may overtly rebel and yet remain dependent. It may feel that it "belongs" by completely conforming to the given cultural patterns at the expense of the loss of its individual self—the result is always a lesser or greater degree of inner emptiness, the feeling of nothingness, anxiety and resulting from all that a chronic hatred, and ressentiment, which Nietzsche characterized very well as Lebensneid, envy of life.[18]

Fromm pointed out that lebensneid is not the same thing as hatred. Hatred has the end goal of the destruction of the object of said hatred. Instead, lebensneid is much more deeply rooted and has the end goal of destroying an object that the subject itself is incapable of enjoying so that nobody else can enjoy it. In this psychological analysis offered up by Fromm and Nietzsche, hatred would be a homicidal impulse and lebensneid would be genocidal impulse, because this envy of life is so perverse as to strive for an end to living itself.

The second connection between ressentiment and lebensneid is metaphorical. In *Thus Spoke Zarathustra*, one of the early manifestations of ressentiment is the fire-dog. But at the same time the fire-dog represents humanity's relation to nature.[19] Here is what Zarathustra has to tell his followers about his conversation with the fire-dog:

> The earth (he [Zarathustra] said) has a skin; and this skin has diseases. One of these diseases, for example, is called "Man." And another of these diseases is called "the fire-dog": men have told many lies and been told many lies about him. . . . At the best, I hold you to be the earth's ventriloquist: and when I have heard subversive and revolutionary devils speak I have always found them like you, bitter, lying and superficial. . . . For the state wants to be absolutely the most important beast on earth; and it is believed to be so, too! When I said that, the fire-dog acted as if he were mad with envy.[20]

The fire-dog epitomizes envy and ressentiment, yet is also the symbol of humanity's relation to nature. We are jealous of the Earth and thus try to create a State and various cultural institutions that we claim to be "better" than nature. Even our architecture often mimics the natural landscapes of nature. This subconscious coveting of nature and wanting to be superior and to control nature causes our inability to enjoy nature and thus our desire to kill nature so that nobody else can enjoy it either. It is amazing that Nietzsche's metaphorical tale is becoming a reality a century later. For the past century, we have been trying to control nature and tame the wilderness and, hence, in our failure to do so we have begun destroying nature through deforestation and pollution. Unfortunately, the irony is that if we destroy nature we will destroy ourselves as well. Lebensneid is also a suicidal impulse when carried out with finality in nature.

Returning to our question about what type of relationship do we have with nature, the answer is one of both biophilia and lebensneid. We have deep-rooted feelings and a natural connection to life; and as the source of this life, wilderness and nature. Yet at the same time, we harbor feelings of resentment and envy towards nature. How else can we explain these dual, opposing perspectives contained in our practice to pollute and poison the very local streams we need and use for drinking water? Children have the same complex relation with nature. Nature is simultaneously a place of both pleasure and pain. For the child, the wilderness holds both the awe of the gentle butterfly and the annoying mosquito waiting to be crushed.

PHILOSOPHICAL REFORMULATIONS: A RECONCEPTUALIZATION OF THE LIFE-WORLD

I have been hesitant to adopt E. O. Wilson's term biophilia fullheartedly. I want to avoid all the genetic and biological determinist talk that goes with Wilson's understanding of biophilia and instead focus on what the meaning of biophilia is. However, like Peter Kahn,[21] I wish to invoke biophilia in a less mechanical, deterministic, and genetic manner. If biophilia is merely a matter of genetics, then what does it matter for us to actively engage the bond if it is merely a case of hardwiring? And a genetic hardwiring, no matter how hard the sociobiologists may try, can never have any direct axiological implications. If this genetic approach to biophilia was analyzed using Kenneth Burke's dialectical rhetorical method of cluster analysis, the genetic would be in direct opposition to the "God terms" of free will, nurture, and environment. Burke's cluster method is valuable because it relies on the Greek term "agon" to look at oppositions in key rhetorical terms. Wilson calls for an ethical examination of the ramification of biophilia, but if biophilia can be collapsed to mere biological determinism, then it has no ethical dimension because it lacks the existential freedom needed for an ethical evaluation.

Perhaps the way to philosophically proceed is with a Husserlian notion of the life-world combined with a rehabilitated notion of Wilson's "biophilia" minus all the baggage of his biological determinism. After a second (or third) *epochē* is performed on the phenomenological level, the seemingly completely solipsistic ego discovers a living, breathing world that it has always been a part of through analogical apperception and horizontalization. For Husserl, it was possible to perform this analogical apperception with animals.[22] Philosophy is in need of a biophilic turn. Since Thales, philosophers sought inspiration from the Earth and the elements around them. Then as we became more and more "civilized," the history of philosophy dealt more and more with human individuality, rationality, and achievement. With an exponential loss of biodiversity and extinction on the planet, many ecologists say we are facing "the death of birth." Meanwhile, even those "environmentally friendly" philosophers have discussions of animals from a "rights based"[23] human perspective, and view the environment and argue for its preservation based almost solely upon its utility. The biophilic turn I'm calling for would not eschew completely these anthropocentric forms of reasoning, but instead call for them to be complemented with a return focus to the life-world that we are all a part of. We need to find that "oceanic feeling."

And the best place to start is in the earliest experience of the child with the wilderness. As we destroy more and more wild spaces, what are we losing? And I don't mean merely the devastating effects to the environment that will affect humanity's health. We need to understand that it is life itself that we are killing. It seems very simplistic to say that part of a definition of being human means to be alive. But this wonderment at "living" itself is part of the infant's gaze and something that we need to recapture. A biophilic focus will be a return to the living and breathing world around us.

However, "biophilia" as a term may be damaged by the genetic and biological determinism Wilson and his genetic and evolutionary biology have attached to it. Therefore, I recommend that a more useful term with the same goal may be the technical phenomenological term "life-world." Now, Husserl's use of this term predates the active modern conservation movements. However, having lived through the horrible technological destruction humanity brought upon itself in World War I,[24] his goal was to save science and technology by showing them as part of the living, breathing world and the impossibility that they could become detached from said world.

Husserl's Originary Ark essay also revealed an understanding, similar to Nietzsche's, of the élan vital of the Earth. The concept of the life-world, first of all, places emphasis on the active processes of a living, breathing biosphere. Husserl would not disagree with the modern Gaia movement that the Earth literally is alive and his concept of the life-world as final and most important step of his phenomenological method reveals this. The life-world is a return to the world that has always been there and made possible our phenomenological analysis in the first place. And as a living, breathing species-being, the Earth is a sphere that we have an intimate connection to as literal and metaphorical mother.

Husserl's concept of the life-world does the work of biophilia without the baggage. If Wilson defined biophilia as "the innate tendency to focus on life and lifelike processes,"[25] my addition or refocus of life-world would allow us to define it as the tendency for the living world to be intimately connected and focused on the living, breathing world and the living plants and animals on its surface. Life has a gravity that attracts other life forms.[26]

So Husserl's conceptualization of the life-world is not contradictory to this conservation element that I want to pull out. I grant that this was not as much of a focus of his writings, but it is part of the term that needs to be pulled out and developed by a new generation of phenomenologists seeking to keep Husserl's phenomenological method alive and relevant. If there is to be a complete bracketing of prejudice towards the world, then we must perform an additional *epochē* that brackets our intimate connection to the entire living world around us. The return to the world via the life-world is a fresh new world that understands the gravity of our connection to life itself and does not take this connection for granted as biophilia does, but instead, conceptualizes it as something that must be nurtured. Our life-world connection to the world around us seems to be slipping away as the forces of lebensneid win out as we operate in an alienated world.

The life-world is also explicitly ethical.[27] Recognizing the world around us as life-world involves both a phenomenological and ethical move. As literally living, the Earth is something that can die. We cannot take the life-world for granted. The life-world implies a responsibility. When we return to an examination of the world as living world of connectivity, we must nurture it, because as life-world it has the ability to die. And sadly, killing the life-world is exactly what we are doing. The planet is very resilient, but we are pushing it to its limits and I worry about the kind of planet we will leave our children. The irreparable harm we are doing to the web of life may be the greatest sin that our children hold us responsible for. I personally worry about the condition of the life-world that I will leave my children. Yet, the first step in repairing the damage that my generation and the generations past have done to the planet may be connecting our children to the world in a way that helps them understand that it is living. Through a child's relationship with animals, a first specific connection to the life of a nonhuman Other is nurtured, and subsequently, a connection to the planet itself can be fostered. Living sustainably and guarding against making too large of a footprint upon the planet *must be* taught to our children.

NOTES

1. World Zoo Organization, *The world zoo conservation strategy executive summary: The role of the zoos and aquaria of the world in global conservation* (Brookfield, IL: Chicago Zoological Society, 1993). www.brookfieldzoo.org/pagegen/inc/wczs.pdf.
2. Robert Mullan, and Garry Marvin. *Zoo culture* (Chicago: University of Illinois Press, 1999), 107.
3. Ibid., 48.
4. Mark Rosenthal, Carol Tauber, and Edward Uhlir. *The ark in the park: The story of Lincoln Park Zoo* (Chicago: University of Illinois Press, 2003), 180.
5. David Samuels. "Wild things: Animal nature, human racism, and the future of zoos." Harper (June 2012): 29–42.
6. Ibid., 36.
7. Philips Verner Bradford, and Ota Blume, *Ota Benga: The pygmy in the zoo* (New York: St. Martins Press, 1992).
8. Edward O. Wilson, *Biophilia* (Cambridge, MA: Harvard University Press, 1984), 1.
9. Edward O. Wilson, *The diversity of life* (New York: W. W. Norton and Company, 1999), 350–51. Also defined biophilia as: "It is not so difficult to love nonhuman life, if gifted with knowledge about it. The capacity, even the proneness to do so, may well be one of the human instincts. The phenomenon has been called biophilia, defined as the innate tendency to focus upon life and lifelike forms, and in some instances to affiliate with them emotionally. Human beings sharply distinguish the living from the inanimate. We esteem novelty and diversity in other organisms." From Edward O. Wilson, *The future of life* (New York: Vintage Books, 2002), 134. Also defined as: "Biophilia, if it exists, and I believe it exists, is the innately emotional affiliation of human beings to other living organisms. From the scant evidence concerning its nature, biophilia is not a single instinct but a complex of learning rules that can be teased apart and analyzed individually." From Edward O. Wilson, "Biophilia and the environmental ethic," in *In search of nature* (Washington, DC: Island Press, 1996), 165.
10. Edward O. Wilson, *Consilience: The unity of knowledge* (New York: Vintage Books, 1999), 12. His exact quotation is, "Philosophy, the contemplation of the unknown, is a shrink- . We have the common goal of turning as much philosophy as possible into

, *In search of nature*.

12. Fischer in *Science*, as quoted in Peter Kahn, *The human relationship with nature: Development and culture* (Cambridge, MA: MIT Press, 1994), 25.

13. Edward O. Wilson, "Biophilia and the conservation ethic," in *The biophilia hypothesis*, ed. Stephen Kellert and Edward Wilson (Washington, DC: Island Press, 1995), 31.

14. Wilson uses the snake as an exemplar in his 1984 *Biophilia* as well.

15. Holmes Rolston, "Biophilia, selfish genes, shared values," in *The biophilia hypothesis*, ed. Stephen Kellert and Edward Wilson (Washington, DC: Island Press, 1995), 381–413.

16. Gilles Deleuze, *Nietzsche and philosophy* (New York: Columbia University Press, 1983), 111.

17. Ibid., 117–18.

18. Erich Fromm, "Selfishness and self-love," in *The yearbook of the International Erich Fromm Society* 5 (2001): 173–97. www.fromm-gesellschaft.eu/index.php/en/publications/on-this-site/yearbook/197-jahrbuch-band-5-1994.

19. Deleuze, *Nietzsche and philosophy*, 139.

20. Friedrich Nietzsche, *Thus spoke Zarathustra: A book for everyone and no one*, trans. R. J. Hollingdale (New York: Penguin, 1981).

21. Peter Kahn, *The human relationship with nature: Development and culture* (Cambridge, MA: MIT Press, 1994), 25.

22. See Edmund Husserl, "Foundational investigations of the phenomenological origin of the spatiality of nature: The originary ark, the Earth, does not move," in Maurice Merleau-Ponty, *Husserl at the limits of phenomenology: Including texts by Edmund Husserl* (Evanston, IL: Northwestern University Press, 2002). Remember he does it with a bird.

23. Such as Peter Singer.

24. It even cost Husserl his son.

25. Edward O. Wilson, *Biophilia* (Cambridge, MA: Harvard University Press, 1984), 1.

26. This even explains our exploration of space. One of our primary missions of space exploration is to seek out new forms of life. Hence, the fascination with possible water sources on Mars and the land rover's collection of soil to look for early forms of life.

27. This is one of the reasons for using phenomenology as a method of investigation for a child's connection with nature. Our relationship with nature can no longer be treated as neutral; we must fully understand the implications of the world as living, giving us a radical existential responsibility.

Epilogue

The purpose of this book was to engage in a phenomenological investigation of the communicative experience of going to the zoo. This book attempted to apply my communicative form of praxis phenomenology. I see it following directly in the footsteps of the phenomenological tradition begun by Edmund Husserl. The phenomenologist is goaded by the call "to the things themselves." In the phenomenological method, we bracket our previous preconceptions and attempt to look at the object of experience in a new light that opens up new meanings. This book attempted to do that by exposing the communicative messages of the zoo experience to the phenomenological gaze.

Zoos are an interesting phenomenon that we visit at an amazing rate. Yet, they have escaped the phenomenological purview despite the interesting challenges they present because of their history, and because their existence on the surface seems to keep animals prisoners solely for human entertainment. As we have seen with the case of people like Ota Benga, zoos have served as actual prisons for some humans as well. Our experiences in the zoo often take place while we are quite young and influence the future bonds we have with other animals, the environment, and even other people. By looking at this particular communicative context, I hope that I have shed some light on the often taken-for-granted experience of the zoo visit.

This book represents a Husserlian phenomenological examination of the zoo experience through the use of a praxis phenomenology. That is not to say that other approaches would not be fruitful, or even offer similar readings. For example, I think of all the work that Gregory Bateson has done on ecology and the attention he paid to communication and systems, and wonder if some of the same conclusions about the nature of the zoo experience would not have been reached by following a Batesonian approach. However, the answer to that question is for another project. But both thinkers would see the zoo as a cultural link between the human and animal world. It is an everyday space that lays open the manifold possibilities of encounter.

Zoos have come a long way from the menageries they were a few centuries ago to today's conservation centers. Zoos present an interesting space of analysis for the phenomenologist. The bonds between human and animal are of particular interest to investigate in the context of the zoo because the space of the zoo and its construction strongly influences our understanding of animal others. A large number of children visit zoos and thus they are responsible for helping shape the story of the way we grow up with animals. As children grow up, they understand that the animals are confined in the zoo, but they also learn appreciation and caring through their interaction with these individual animals that serve as a synecdoche for the zoo, the species, and oftentimes, all animals. One of the key characteristics that I observed is the role of mobility and rest involved in children's interaction with these zoo animals. Husserl recognized the importance of movement in his genetic phenomenology and developed a method of analysis called "kinesthesia" that allows for an analysis of movement and how motion affects our understanding and bond with the world. One of the most important eidetic structures this project laid bare is the kinesthetic role of the bond between children and all animals.

In the final chapter I conducted a genetic historical analysis of the evolutionary history of zoos. I highlighted the contentious colonial history of zoos. It is my belief that this history needs to be more honestly and publicly accounted for. I also carried out a generative phenomenology of what it is that drives us to seek out zoos and animals. Biophilia is the belief that we have an innate affinity towards living things. I agree with biophilia's assessment of a connection between human and earth, but disagree that it is an innate hereditary trait stored in our genes. I believe that biophilia is a structure of the life-world that Husserl was writing about a century ago. Phenomenologists should lead the way in the current philosophical investigations of how we experience this living Earth and can provide a valuable tool for the environmental movement by thoroughly exploring the living structure of the life-world.

I leave you with a few takeaways from this analysis for both zoo visitor and zoo director. For the zoo visitor:

1. Let your child play in the zoo. This is an opportunity for developing early forms of empathic thinking with animals. Help them become a turtle, roar like a lion, or leap like a lemur.
2. Just stop and watch the animals. Teach your children how to be in awe of animals. Use the zoo to model animal appreciation.

3. Teach your children about the fragility of ecosystems and how we humans are radically impacting the animal world with our actions. Then teach your children how they can live a more sustainable existence that will help protect the wild homes of these zoo animals.
4. Don't avoid the paradoxes. Children are much more perceptive then you realize. Talk to them about the phenomenon of captivity and freedom in the context of the zoo. Use the zoo to discuss the wild and the endangered places around the world.
5. Watch your child in the zoo. It is not a playground. You need to be a parent and make sure they understand what they are seeing and are not endangering themselves or causing extra stress to the animals.
6. Finally, teach your child compassion. The most important thing you can do for the future of the animal world is to raise children who care. Teach them to care *about* animals, and care *for* the earth.

For the zoo director:

1. Talk about the history of the zoo. From an ethical perspective it is important for zoos to come clean about their troubled past practices, particularly in their treatment of various indigenous peoples. These origins are often carried in our collective unconscious and telling these tales helps the zoo differentiate themselves from their former manifestations.
2. Involve the public more actively in conversations about the care of animals kept at the zoo. Let people know where animals came from and are going. Talk with the public about perceptions of animal happiness. Involve the public in the decision-making process of what animals should be kept in what spaces.
3. I believe that people want to do good, but often don't know how. It is not enough for zoos to sound the alarm on issues like climate change, but they need to provide accurate information of behaviors that visitors can take to help in the conservation fight.
4. Continue and expand your role as researchers of both animals and ecosystem management.
5. Become more active advocates of change. It is time for the voice of zoos to speak out with an even louder voice for conservation. Zoos need to be more vocal in the public sphere.

I began this praxis phenomenological examination with my story of going to the zoo as a child with my mother. These experiences were important in shaping my understanding of animals and our relationship to others in general. Now, as a father myself, I continue this tradition of taking my children to different zoos around the world and see the experience of the zoo visit with a new set of eyes. I wonder if their understandings of the zoo and constituted meanings will be the same as the ones I had as a child? Zoos are changing, and rightfully need to change, to stay relevant to our contemporary world, but they must always be aware of their contested and troublesome origins. I wonder what the zoos will look like for the next generation as more animals become extinct and we become increasingly detached from the lived earth. One thing I am certain of is that the stories of the zoo visit will continue to play an important formative role in our relationship with the animal world.

Bibliography

Aristotle. *Rhetoric.* Translated by George A. Kennedy. New York: Oxford University Press, 1991.

Association of Zoos and Aquariums. "Public benefits of AZA." www.aza.org/public-benefits/.

Beauvoir, Simone de. *The prime of life.* Chicago, IL: Paragon, 1992.

Beck, Alan, and Aaron Katcher. Between pets and people. West Lafayette, IN: Purdue University Press, 1996.

———. "Future directions in human-animal bond research." *American Behavioral Scientist.* 47. no. 1 (2003): 79–93. doi: 10.1177/0002764203255214.

Blake, William. "The Tyger." In Songs of experience. Mineola, NY: Dover Publications, 1984.

Bradford, Philips Verner, and Ota Blume. Ota Benga: The pygmy in the zoo. New York: St. Martins Press, 1992.

Bradley, Ed. Tiger Woods television interview, March 26, 2006.

Chan-Fai, Cheung, Ivan Chvatik, Ion Copoeru, Lester Embree, Julia Iribarne, and Hans Rainer Sepp, eds. "Essays in celebration of the founding of the Organization of Phenomenological Organizations." 2003. www.o-p-o.net.

Churchman, David. "How and what recreational visitors learn at zoos." Paper presented at Annual Western Meeting of the American Association of Zoological Park and Aquarium Administrators, Anchorage, Alaska, March, 1985.

Clubb, Ros, and Georgia Mason. "Animal welfare: Captivity effects on wide-ranging carnivores." *Nature* 425 (2003): 473–74.

Clutton-Brock, Juliet. "Aristotle, the scale of nature, and modern attitudes to animals." In *Humans and other animals*, edited by Arien Mack, Chapter 1. Columbus: Ohio State University Press, 1995.

Connolly, Kate. "Rejected at birth, Knut becomes Berlin zoo's bear essentials." The Guardian , March 23, 2007. www.guardian.co.uk/world/2007/mar/24/animalwelfare.germany.

Cox, Robert. "The die is cast: Topical and ontological dimensions of the locus of the irreparable." Quarterly Journal of Speech . 68. (1982): 227–39.

———. Environmental communication and the public sphere . Los Angeles, CA: Sage Publication, 2009.

Crossland, David. "Farewell Knut, thanks for the memories." Der Spiegel , sec. Obituary, March 21, 2011. www.spiegel.de/international/germany/obituary-farewell-knut-thanks-for-the-memories-a-752247.html.

Crowell, Galt. "Does the Husserl/Heidegger feud rest on a mistake? An essay on psychological and transcendental phenomenology." Husserl Studies 18, no. 2 (2002): 123–40. doi: 10.1023/A:1015540519310.

De Castella, Tom. "Knut: A polar bear story." BBC News , March 22, 2011. www.bbc.co.uk/news/magazine-12805534.

De Waal, Frans. Our inner ape . New York: Riverhead Books, 2005.

Deleuze, Gilles. Nietzsche and philosophy . New York: Columbia University Press, 1983.

Derrida, Jacques. The animal that therefore I am . New York: Fordham University Press, 2008.

———. "The animal that therefore I am (More to follow)." Critical Inquiry 28, no. 2 (2002): 415–16. www.jstor.org/stable/1344276.

Embree, Lester. "The natural-scientific constitutive phenomenological psychology of humans in the earliest Sartre." Research in Phenomenology 9, no. 11 (1981): 41.

Farrell, Thomas, and Thomas Goodnight. "Accidental rhetoric: The root metaphors of Three Mile Island. "Communication Monographs . 48: 271–300. doi: 10.1080/03637758109376063

Foucalt, Michel. Order of things: An archeology of the human sciences . New York: Vintage Press, 1994.

Fromm, Erich. "Selfishness and self-love." In *The yearbook of the International Erich Fromm Society*. 5 (2001): 173–97, www.fromm-gesellschaft.eu/index.php/en/publications/on-this-site/yearbook/197-jahrbuch-band-5-1994.

Geertz, Clifford. The interpretation of cultures: Selected essays . New York: Basic Books, 1973.

Glickman, Stephen. "The spotted hyena from Aristotle to the Lion King: Reputation is everything." In *Humans and other animals*, edited by Arien Mack. Columbus: Ohio State University Press, 1995.

Grandin, Temple and Catherine Johnson. Animals in translation . Orlando, FL: Harcourt Inc, 2005.

Guillaume, Paul. Imitation in children . Chicago, IL: University of Chicago Press, 1971.

Gurwitsch, Aron. Studies in phenomenology and psychology . Evanston, IL: Northwestern University Press, 1966.

Haar, Michael. The song of the Earth: Heidegger and the grounds of the history of being . Bloomington: Indiana University Press, 1993.

Hanson, Elizabeth. Animal attractions: Nature on display in American zoos . Princeton, NJ: Princeton University, 2002.

Heidegger, Martin. Being and time . Translated by John Macquarrie and Edward Robinson. New York: Harpers, 2008.

Hopkins, Burt. "The Husserl-Heidegger confrontation and the essential possibility of phenomenology: Edmund Husserl, psychological and transcendental phenomenology and the confrontation with Heidegger." Husserl Studies 17, no. 2 (2001): 125–48. doi: 10.1023/A:1010771906591.

Huizinga, Johan. Homo ludens: A study of the play element in culture . Boston, MA: Beacon Press, 1960.

Husserl, Edmund. Analyses concerning passive and active synthesis: Lectures on Transcendental Logic . Translated by Anthony Steinbock. Boston, MA: Kluwer Academic Publishers, 2001.

———. *Cartesian meditations*. Translated by Dorion Cairns. The Hague: Springer Press.1967.

———. The crisis of European sciences and transcendental phenomenology . Evanston, IL: Northwestern University Press, 1997.

———. Formal and transcendental logic . The Hague: Nijhoff, 1969.

———. "Foundational investigations of the phenomenological origin of the spatiality of nature: The originary ark, the Earth, does not move." In Maurice Merleau-Ponty. Husserl at the limits of phenomenology: Including texts by Edmund Husserl . Evanston, IL: Northwestern University Press, 2002.

———. Ideas pertaining to a pure phenomenology and to a phenomenological philosophy: First book . Boston, MA: Kluwer Academic Publishers, 1989.

———. Ideas pertaining to a pure phenomenology and to a phenomenological philosophy: Second book . Boston, MA: Nijhoff, 1989.

———. Ideas pertaining to a pure phenomenology and to a phenomenological philosophy: Third book . Boston, MA: Nijhoff, 1980.

———. Logical investigations, Vol. 2. New York: Humanities Press, 1970.

———. "Phenomenology." In The encyclopedia britannica . 4 . Vol. 1. London: Encyclopedia Britannica Inc, 2004.

———. Psychological and transcendental phenomenology and the confrontation with Heidegger (1927–1931) Translated and edited by Thomas Sheehan and Richard Palmer. Dordrecht: Kluwer Academic Publishers, 1997.

———. "Pure phenomenology, its method and its field of investigation." Presented at Inaugural Freiburg Lecture, 1917. www3.baylor.edu/~Scott_Moore/essays/Husserl.html.

Hyson, Jeffery. "Education, entertainment, and institutional identity at the zoo." Curator: The Museum Journal . 47. no. 3 (2004): 247–51. doi: 10.1111/j.2151–6952.2004.tb00122.x.

Irigaray, Luce. "Animal compassion." In *Animal philosophy: Ethics and iIdentity*. Edited by Matthew Calarco and Peter Atterton. Translated by Marilyn Gaddis Rose. London: Continuum, 2004, 195–201.

Jamieson, Dale. "Against zoos." In In defense of animals . Edited by Peter Singer. Oxford: Blackwell, 1985.

Kahn, Peter. The human relationship with nature: Development and culture . Cambridge, MA: MIT Press, 1994.

Katcher, Aaron, Alan Beck, Erika Friedmann, and James Lynch. "Looking, talking and blood pressure: The physiological consequence of interacting with the living environment." In *New perspectives on our lives with companion animals*. Edited by Aaron Katcher and Alan Beck. Philadelphia: University of Pennsylvania Press, 1996, 351–59.

Lauer, Quentin. "Questioning the phenomenologists." Journal of Philosophy 58, no. 21 (1961): 633.

Lawlor, Leonard. This is not sufficient: An essay on animality and human nature in Derrida . New York: Columbia University Press, 2007.

Leibovitz, Annie. "Leo and the bear." Vanity Fair , April 2, 2007. www.vanityfair.com/politics/features/2007/05/knut_slideshow200705.

Levinson, Boris. "Interpersonal relationships between pets and human beings." In Abnormal behavior in animals . Edited by Michael Fox. Baltimore, MD: Williams and Wilkins, 1975.

———. "Pets, child development and mental illness." Journal of the American Veterinary Medical Association 157, no. 11 (1970): 1759–766.

Lewis, Gordon. Her majesty's other children: Sketches of racism from a neocolonial age . Lanham, MD: Rowman & Littlefield Publishers, Inc., 1997.

Louv, Richard. Last child in the woods . Chapel Hill, NC: Algonquin Books, 2006.

Malamud, Randy. Reading zoos: Representations of animals and captivity . New York: New York University Press, 1998.

Maryberry, Sara. Innovations in healthcare design . New York: Van Nostrand Reinhold, 1995.

Mason, Georgia, and Jeffery Rushen. Stereotypic animal behavior: Fundamentals and applications to welfare . Cambridge, MA: Cabi, 2006.

McBride, William. "John Wild and the life-world." In *Analecta Husserliana, vol. XXVI*. Edited by Eugene F. Kaelin and Calvin O. Shrag. Secaucus, NJ: Kluwer Academic Publishers, 1989.

McCormick, Peter, and Elliston Frederick, eds. Husserl: Shorter works. Translated by Robert W. Jordan. Introduced by H. L. Van Breda. Notre Dame, IN: University of Notre Dame Press, 1981, 9–17.

Mead, Margaret. "Cultural factors in the cause and prevention of pathological homicide." *Bulletin in the* Menninger Clinic . 28. (1964): 11–22.

Merleau-Ponty, Maurice. Nature: Course notes from the College de France . Evanston, IL: Northwestern University Press, 2003.

———. Phenomenology of perception . New York: Routledge, 1999.

Mezei, Balazs M. World and life-world aspects of the philosophy of Edmund Husserl . New York: European University Studies Press, 1995.

Moran, Dermot. Introduction to phenomenology . New York: Routledge Press, 2000.

Mullan, Robert and Garry Marvin. Zoo culture . Chicago: University of Illinois Press, 1999. www.waza.org/en/site/about-waza/who-we-are.

Myers, Olin, Carol Saunders, and Andrej Birjulin. "Emotional dimensions of watching zoo animals: An experience sampling study building on insights from psychology." Curator: The Museum Journal 47, no. 3 (2004): 299–321. doi: 10.1111/j.2151–6952.2004.tb00127.x.

Myers, Olin E., Carol D. Saunders, and Erik Garrett. "What do children think animals need? Aesthetic and psycho-social conceptions." Environmental Education Research 9, no. 2 (2003): 305–25. doi: 10.1080/13504620303461.

———. "What do children think animals need? Developmental trends." Environmental Education Research 10, no. 4 (2004): 545–62. doi: 10.1080/1350462042000291056.

Nagel, Thomas. "What is it like to be a bat?." The Philosophic Review LXXXIII (1974): 435–50. http://rintintin.colorado.edu/~vancecd/phil1000/Nagel.pdf.

Nietzsche, Friedrich. Thus spoke Zarathustra . Translated by Walter Kaufmann. New York: Penguin, 1980.

———. Thus spoke Zarathustra: A book for everyone and no one . Translated by R. J. Hollingdale. New York: Penguin, 1981.

Nissim-Sabat, Marilyn, and David Fryer. "Mission statement of the Phenomenology Roundtable." Published as essay 38 in Cheung Chan-Fai, Ivan Chvatik, Ion Copoeru, Lester Embree, Julia Iribarne, and Hans Rainer Sepp, eds. "Essays in celebration of the founding of the Organization of Phenomenological Organizations." 2003. www.o-p-o.net.

Ohio Department of Natural Resources, "Eastern ratsnake or black ratsnake." www.dnr.state.oh.us/Home/species_a_to_z/SpeciesGuideIndex/blackeasternratsnak e/tabid/6556/Default.aspx.

Oravec, Christine, "Conservationism vs. preservationism: the 'public interest' in the Hetch Hetchy controversy." Quarterly Journal of Speech 70: 444–58. doi:10.1080/00335638409383709.

Piaget, Jean. Play, dreams, and imagination in childhood . New York: Norton, 1962.

"Polar bear controversy." Reported by Richard Schlesinger. CBSnews. March 22, 2007. www.cbsnews.com/video/watch/?id=2599094n.

Power, Thomas. Play and exploration in children and animals . Mahwah, NJ: Lawrence Erlbaum and Associates, 2000.

Rabb, George. "The evolution of zoos from menageries to centers of conservation and caring." Curator: The Museum Journal 47, no. 3 (2004): 237–46. doi: 10.1111/j.2151–6952.2004.tb00121.x.

Regan, Tom. The case for animal rights . Los Angeles: University of California Press, 1983.

Rollin, Bernard. "Through a frame darkly: A phenomenological critique of zoos." In *Metamorphoses of the zoo: Animal encounter after Noah*. Edited by Ralph R. Acampora. Boulder, CO: Lexington Books, 2010.

Rolston, Holmes, "Biophilia, selfish genes, shared values." In The biophilia hypothesis . Edited by Stephen Kellert and Edward Wilson. Washington, DC: Island Press, 1995.

Rosenthal, Mark, Carol Tauber, and Edward Uhlir. The ark in the park: The story of Lincoln Park Zoo . Chicago: University of Illinois Press, 2003.

Samuels, David. "Wild things: Animal nature, human racism, and the future of zoos." Harper , June 2012, 29–42.

Seidman, Robert. "Discovery Channel's 23rd annual 'Shark Week' is most-watched ever; Seen by 30.8 million people." *TV by the Numbers*. Last modified 2010. http://tvbythenumbers.zap2it.com/2010/08/10/discovery-channels-23rd-annual-shark-week-is-most-watched-ever-seen-by-30-8-million-people/59634/.

Shrag, Calvin. Radical reflection and the origin of the human sciences . West Lafayette, IN: Purdue University Press, 1980.

Simpson, George Gaylord. *Principles of animal taxonomy.* As quoted in Juliet Clutton-Brock. "Aristotle, the scale of nature, and modern attitudes to animals." In *Humans and other animals*, ed. Arien Mack, (Columbus: Ohio State University Press, 1995), chap. 1.

Singer, Peter. Animal liberation: A new ethics for our treatment of animals . New York: New York Review and Random House, 1975.

Soloman, Robert. Phenomenology and Existentialism . New York: University Press of America, 1980.

Spotte, Stephen. Zoos in postmodernism: Signs and simulation . Madison, NJ: Fairleigh Dickinson University Press, 2006.

Theunissen, Michael. The other: Studies in the social ontology of Husserl, Heidegger, Sartre, and Buber . Cambridge, MA: MIT Press, 1984.

Ulrich, Roger. "Effects of health facility interior design on wellness: Theory and scientific research." Journal of Health Care Design . 3. (1991): 97–109.

———. "View through a window may influence recovery from surgery." Science 224 (1984): 420–21. www.ncbi.nlm.nih.gov/pubmed/6143402.

Ulrich, Roger, Outi Lundén, and John Eltinge. "Effects of exposure to nature and abstract pictures on patients recovering from heart surgery." Paper presented at the Thirty-Third Meetings of the Society for Psychophysiological Research, Rottach-Egern, Germany. 1993. Abstract published in *Psychophysiology* 30 (Supplement 1, 1993): 7.

Ulrich, Roger, Robert Simons, Barbara Losito, Evelyn Fiorito, Mark Miles, and Michael Zelson. "Stress recovery during exposure to natural and urban environments." Journal of Environmental Psychology 11 (1991): 201–30.

Velarde-Mayol, Victor. On Husserl . Belmont, CA: Wadsworth, 2000.

Waddell, Craig. Landmark essays: On rhetoric and the environment . Mahwah, NJ: Lawrence Erlbaum and Associates, 2007.

Warner, Mary Beth. "Polar bear will be given home at natural history museum." Der Spiegel , March 25, 2011. www.spiegel.de/international/germany/remembering-knut-polar-bear-will-be-given-home-at-natural-history-museum-a-753142.html.

Wild, John. Human freedom and social order: An essay in Christian philosophy . Durham: Duke University Press, 1959.

Wilson, Edward O, "Biophilia and the conservation ethic." In The biophilia hypothesis , edited by Stephen Kellert and Edward Wilson. Washington, DC: Island Press, 1995.

———. Biophilia . Cambridge, MA: Harvard University Press, 1984.

———. *Consilience: The unity of knowledge*. New York: Vintage Books, 1999.

———. The diversity of life . New York: W. W. Norton and Company, 1999.

———. The future of life . New York: Vintage Books, 2002.

———. In search of nature . Washington, DC: Island Press, 1996.

World Zoo Organization. The world zoo conservation strategy executive summary: The role of the zoos and aquaria of the world in global conservation . Brookfield, IL: Chicago Zoological Society, 1993. www.brookfieldzoo.org/pagegen/inc/wczs.pdf.

Zahavi, Dan. Husserl's phenomenology . Stanford, CA: Stanford University Press, 2003.

Index

About the Author

Erik A. Garrett is associate professor of communication and rhetorical studies at Duquesne University. He earned his PhD from Purdue University in both philosophy and communication. He is a cofounder of the Phenomenology Roundtable and is active in several philosophy, communication, and interdisciplinary journals and organizations.